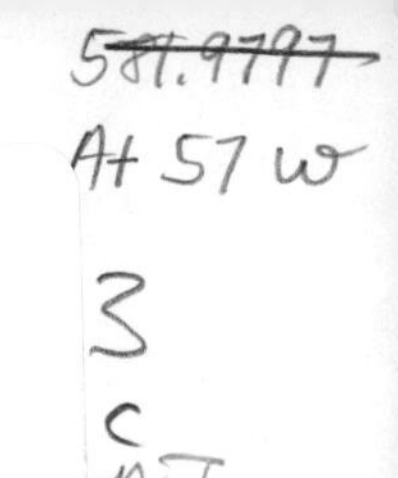

WILD PLANTS of the San Juan Islands

Scott Atkinson & Fred Sharpe

THE MOUNTAINEERS

The Mountaineers: Organized 1906
"... to explore, study, preserve, and enjoy
the natural beauty of the Northwest."

Published by The Mountaineers
306 Second Avenue West, Seattle, Washington 98119

Published simultaneously in Canada by Douglas & McIntyre, Ltd.
1615 Venables Street, Vancouver, British Columbia V5L 2H1

Manufactured in the United States of America

Edited by Barbara Chasan
Designed by Constance Bollen
Illustrations and cover photo by Fred Sharpe
Cover photo: Few-flowered shooting star *(Dodecatheon pulchellum)*

Library of Congress Cataloging in Publication Data

Atkinson, Scott, 1959-
Wild plants of the San Juan Islands.

Bibliography: p.
Includes index.
1. Botany—Washington (State)—San Juan Islands.
2. Plants—Identification I. Sharpe, Fred. II. Title.
QK192.A85 1985 581.9797'74 85-10661
ISBN 0-89886-104-7

DEDICATION

We dedicate this book to our parents, who planted the seeds of curiosity that have grown into an integral part of our lives.

Contents

Acknowledgments

We would first like to thank The Mountaineers for publishing this book for us. Peter Rapp played an inspirational role by encouraging the project at the outset. Joan Davy and Gretchen Junker, by being supportive companions, helpers, and sources of inspiration, just when such persons were needed, were instrumental as well.

Dr. Estella Leopold is thanked for her encouragement and for giving us an excellent place in which to work. We also thank Dr. Arthur Kruckeberg and Dr. Melinda Denton for their words of wisdom and support. We thank William Baker from San Juan Island for his anecdotes, enthusiasm, and valuable information, and Betty Higinbotham for her helpful clues. Also from San Juan we thank Dick and Janet Wright, and Matt Mottolas, for their friendly help. Dr. Frank Richardson and Walter Harm are thanked for their information; Dave Castor, another Orcas Islander, provided boat transportation for several excursions. Emily Reed's kindness was exceeded only by her knowledge and charm, for which we can only say thank you.

Two Canadians, Harvey Janzsen and Dr. Adolf Ceska, revealed to us the superior understanding that many Canadians, particularly those on Vancouver Island and nearby, have of their flora and fauna. We thank Anna Ziegler at the University of Washington herbarium, Dr. Ronald Taylor at Western Washington University, Dr. G.D. Alcorn at the University of Puget Sound, Paul M. Peterson from the Marion Ownbey herbarium at Washington State University, and Kenton L. Chambers at Oregon State University, for their cheerful cooperation. Linda Kunze is thanked for sharing her expertise on wetland and salt-marsh flora. Both the Washington Natural Heritage Program and The Nature Conservancy were helpful, and we are grateful for their interest in the San Juans. Dr. Eugene Kozloff, Tim and Jane Nelson, Dr. Ross Shaw, Wally Lumm, Paul Sharpe, Dr. William and Molly Wolfe, Phyllis Wood, Rhoda Love, Richie Kehl, Ed Alversen, Arthur Jacobsen, and Wenonah Sharpe are also thanked for their generous support.

Finally, we wish to express our sincere gratitude to the staff of The Mountaineers Books. The expertise and understanding provided by Ann Cleeland, Donna DeShazo, and Stephen Whitney, as well as Evelyn Peaslee and other members of the Editorial Review Committee have proven invaluable.

PREFACE

Appealing to resident and visitor alike, the San Juans are well-known to many for their recreational virtues. Bikers, hikers, boaters, and many others come to the islands in droves during summer, causing a burgeoning tourist trade that feeds the islands' economy. Marine attractions seem to have won most of the interest: boaters, scuba divers, fishermen, whale watchers, and researchers all are in some way involved with the islands' marine environment, whether they are out for a sight-seeing trip or trying to catch a glimpse of a pod of Killer whales. However, the islands themselves have also been the object of much enthusiasm, lauded for their undeveloped character, their scenic views, and much more. Only the most inattentive person could fail to notice the hilly, rocky topography which is so different from that of other parts of Puget Sound. Rocky balds and meadows abound; in springtime, they feature a blaze of wildflower color that has no match in Puget Sound country.

Many islanders, of course, have long been aware of the island flora, and their admiration for it is reflected in the many references found in local publications like the *San Juan County Almanac* and in the disproportionately large number of nature lovers one encounters. Wildflower classes have been taught, reports have been made on the flora of several small islands and plots, and many persons simply go on walks to enjoy wildflowers or nature in general. On certain islands the reverence for native flora and fauna is very intense; indeed, a Lopez islander once described wildflower walks by some residents in spring as "almost a ritual." In light of all this, it is nothing short of amazing that no guide to San Juan flora, or at least to common San Juan wildflowers, has been previously published. Probably the difficulty in covering such an inaccessible region has proven an impediment to many.

This work is an attempt to give general information on 192 plant species, including wildflowers, shrubs, a few trees, some weeds, and other plants that occur commonly, for the most part, throughout the archipelago—as well as on the Canadian Gulf Islands, southern Vancouver Island, and on Fidalgo, Whidbey, and other islands to the east of the main San Juans. Common and Latin names are provided for each species, followed by a general description of key field marks, meaning of the name, habitat specifics, and so forth. The accounts are brief (four–ten sentences) and no attempt is made to give exhaustive details on visual plant characteristics, for the simple reason that a number of guides are available in which one can find lengthy descriptions concerning that matter. Within each chapter, plants are arranged in the following order: ferns and fern-allies, grasses and grasslike plants, wildflowers, shrubs, and trees. Wildflower and shrub sections are further organized by color, in the following progression: white, yellow, orange, red, pink, purple, blue, green, and brown.

Chapters themselves are distinguished by habitat, with the twofold understanding that (a) species described under one habitat will often show up in another one, e.g., the intergradation of rocky outcrop and meadow species, and (b) the seven habitats, of which the chapters are comprised, are general in that each of them could be further split into a number of sub-habitats and communities. Preceding the species accounts are habitat descriptions. Habitat and species descriptions are geared to the casual observer, with technical terms generally deleted or else clarified by the plant diagrams (below) or the glossary (p. 132). However, for the more experienced botanists, the last part of the book includes an annotated checklist of all vascular plants recorded for San Juan County, plus an introductory section detailing research history, important contributors to the compilation (i.e., famous collectors and current sources), and associated matters. The book concludes with a bibliography and an index.

We hope that this book will serve as an inspiration for islanders and visitors alike to investigate the San Juans' floral attractions, but we do so with some qualifications. First, we ask that observers not pick wildflowers or other sensitive plants on public lands, but ask that they be left for others to enjoy. Many wildflowers take years to regenerate and, thus, random picking can quickly decimate a population. Second, if and when collecting wild edibles, we encourage individuals to do so only on their own land (except with the most ubiquitous species, e.g., blackberries, Wild carrot, etc.) and to be absolutely certain of the identification before consuming. A number of deadly poisonous plants occur in the San Juans, and several of them bear likenesses to desirable edible species. We also request that private property boundaries always be respected; trespassing is a crime. Finally, if you have bought this book, or only borrowed it, thank you.

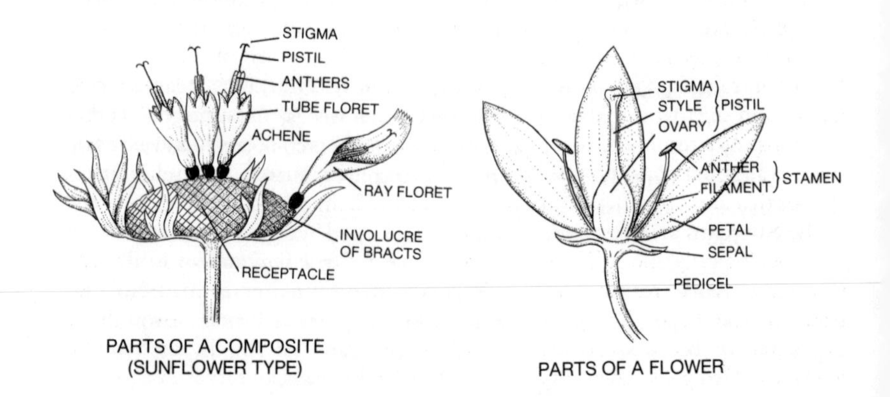

PARTS OF A COMPOSITE (SUNFLOWER TYPE)

PARTS OF A FLOWER

INTRODUCTION

The San Juan Islands proper consist of 473 islands, the largest of which are Orcas (57 square miles), San Juan (55 square miles), and Lopez (29 square miles). Other important islands include Stuart and Waldron at the northwest corner of the chain; the Sucias, Patos, and Matia at the northern edge; Decatur and Blakely to the east; and Shaw and the Wasp–Yellow Island group in the interior. The highest peaks are Mt. Constitution (2409 feet), Mt. Pickett (1765 feet), and Mt. Woolard (1180 feet), on Orcas Island; Mt. Dallas (1036 feet) on San Juan Island; and Blakely Peak (1042 feet) on Blakely Island.

Commonly mentioned places on San Juan Island in this book include Cattle Point–South Beach at the southeast corner, Deadman Bay along the west–central coast, Sportsman–Egg Lakes in the interior, and the towns of Friday Harbor (1983 estimated year-round population: 1200) and Roche Harbor near the northwest corner.

On Lopez, important points include Iceberg Point–Point Colville at the southern edge, Davis Head and Fisherman Bay–Lopez Village on the west–central coast, and Spencer Spit and the Sperry Marsh along the east coast.

On Orcas, the commonly cited places are Mt. Constitution, Summit, Mountain, and Cascade Lakes on the east lobe, the town of Eastsound at the northern tip, and Killebrew Lake near the southern tip of the west lobe.

The San Juans lie within the marine climate belt that covers the Northwest's Pacific Coast. Comparatively cool, dry summers and mild, wet winters are normal, with approximately 70 percent of the precipitation falling between October and March. Extremes in temperature are rare at any season. Low annual precipitation is also characteristic, a direct result of the Olympic Mountain rain shadow; most areas in the San Juans average between 20–30 inches annually, which is significantly lower than that of most Puget Sound locations. As a result, the San Juan Islands contain some species found east of the Cascades, but not often found west of the mountain range. Conversely, some species that are common west of the Cascades and that one would expect to find in the San Juans are mysteriously absent.

Geologically, the San Juans are discordant with the regional topography, forming erosional resistant terrain perpendicular to the general north–south trend of the Puget Trough. As is evident, the islands contain extensive exposures of bedrock, an uncommon development in the Puget Sound lowlands. Two geologic events, the uplift of land masses and the subsequent glaciations, are responsible for the incongruous landscape; today, the islands bear considerable evidence, in particular, of the Vashon glaciation, which ended about 12,000 years ago and was the last of four major glaciations that occurred in the Puget Sound region during the Pleistocene Epoch. Underlying the omnipresent glacial deposits found throughout the archipelago are, in general, young, unmetamorphosed sedimentary materials to the north and older, partially metamorphosed and more complex formations southward. Soils develop principally upon deposits left behind by the receding glaciers. Clay and rock fragments, of sizes varying from tiny pebbles to huge boulders, are common in the coarse-textured San Juan soils.

Coverage area of book

Specific locations within the Islands mentioned in plant descriptions.

1
MEADOWS

Striking to even the most casual of spring visitors, the meadows of the San Juans offer a kaleidoscope of wildflower color not readily matched in Puget Sound country. Several factors contribute to the rich floral diversity found here, including the natural environmental restrictions that have prevented tree growth. South-facing slopes, exposed to greater climatic stress than that found in other places, are the most common meadow sites. Wherever the locale, however, botanic composition strongly resembles that of the open, rocky outcrop, with which the meadows are often associated.

San Juan meadows were originally the result of the Vashon glaciation, which left a thin veneer of silt, sand, and gravel atop the existing rock structure. Here, conditions gradually became suitable for vegetative colonization; floral composition varied through time before reaching its current status. Then and now, environmental constraints—intense summer sunlight, winter winds, and the area's relative lack of precipitation—have prevented vegetation from progressing past the grass–herb stage. Inadequate soil quality and depth are at least as important. The glacial deposits are often coarse grained and sandy and, thus, poor in nutrient and water retention. Heavy winter rains quickly leach out nutrients; sunlight causes dehydration during summer.

Meadows are a noticeable element of the San Juan landscape, evident at such places as Mt. Constitution and the southern slopes of the Turtleback Range on Orcas Island, the west side of San Juan Island, Iceberg Point on Lopez Island, and on Spieden and Yellow Islands. At these places one may find the fullest expression of the San Juan meadow, but whether at one of the named locations or some other meadow, the spring visitor to the San Juans is apt to encounter a wildflower display not soon forgotten.

Sheep sorrel *(Rumex acetosella)* is edible when young and seems to show up in virtually any habitat; however, it is particularly abundant in meadow situations. Brilliant yellow spreads over grassy knolls in spring, and Western buttercup *(Ranunculus occidentalis)* is usually the source. Various diminutive clovers are less apparent, with Tomcat clover *(Trifolium tridentatum)* being perhaps most common. Rich violet clusters are provided by Western long-spurred violet *(Viola adunca)*, especially common on the Cattle Point meadows. Spring gold *(Lomatium utriculatum)*, named appropriately, is scattered among other herbs. Perhaps most appealing of the meadow denizens are the shooting stars *(Dodecatheon pulchellum* and *hendersonii)*. Yarrow *(Achillea millefolium)* grows abundantly here as elsewhere.

Lilies are very well represented, with several species displaying showy blooms in spring. Hooker's onion *(Allium acuminatum)* is abundant, covering whole hillsides with its rosy blooms. Two brodiaeas are worthy of note: Harvest and Hyacinth brodiaea *(Brodiaea coronaria* and *hyacinthina,* respectively). The former seems very common during June in almost any meadow, while the latter, a plant with umbellate white flower clusters, reaches greatest abundance on San Juan Island, where it even grows in roadside ditches. Great camas *(Camassia leichtlinii)* is abundant throughout the archipelago, its purple clusters a splendid sight. Poison-camas *(Zigadenus venenosus),* a showy, white-flowered species, usually grows near Great camas. Brown, nodding heads scattered along a stem means the observer has located Chocolate lily *(Fritillaria lanceolata).* This distinctive lily covers the Iceberg Point meadows with a surprising density during April. Upper, wetter meadows usually have Blue-eyed grass *(Sisyrinchium angustifolium)* in residence. Finally, Crane's bill *(Erodium cicutarium)*, a pink geranium with fernlike leaves, grows commonly on both meadows and open, rocky outcrops.

GRASSES AND GRASSLIKE PLANTS

RED FESCUE *(Festuca rubra)*. From the highest meadows of Mt. Constitution to the sandy beaches of Cattle Point, Red fescue seems to be everywhere. There are a number of equally numerous grasses that are despicable in every respect; however, Red fescue's abundance is outweighed by its handsomeness. The burgundy to brownish basal sheaths of Red fescue, which are strongly nerved and tear in fibers, are most appealing, especially as sunlight strikes them. More striking is the new growth, which exhibits a unique turquoise-bluish tint, especially in var. *littoralis,* our common shoreline form. Var. *rubra* is the common upland type. Idaho fescue *(F. idahoensis)* is similar but has greenish basal sheaths, which do not tear in fibers, and is non-rhizomatous; however, these characteristics are not always reliable, rendering some specimens indistinguishable in the field. Less common; prefers higher elevations, as on Mt. Constitution.

FLOWERS

White

SMALL-FLOWERED PRAIRIE STAR *(Lithophragma parviflora)*. Harmoniously proportioned, each white or light pink star consists of five trilobed petals attached to a small cup. One will often find this humble, well-named wildflower in attendance on the prairies of eastern Washington or amid the early spring bloomers on a moist San Juan meadow. Withering quickly with the departure of spring moisture, plants become rather scraggly and unrecognizable by late May and are gone soon afterward. Leaves are three–five times lobed, and, along with the up to 2-foot stems, are glandular-hairy. Locally common.

OREGON MANROOT *(Marah oreganus)*. Suggestive of cucumber, the woody vines of this gargantuan plant snake across meadows or climb hillsides and onto nearby vegetation. On one occasion as a child one of your authors had the sorry experience of mistaking the spiny, hard melons for some cultivated cucumbers that had gone wild; one bite of the very bitter fruit has proved a sufficient impediment to any future miscalculations. Accordingly, the genus name, *Marah,* Hebrew for bitter, seems to be quite appropriate. Growing to 20 feet long or more, the sprawling vines, with their strangling tendrils and monstrous, dark green leaves, would seem more appropriate in an Amazon jungle. In fact, along the north shore of Grays Harbor, where they form a conspicuous part of the understory, they give the woodland underbrush the look of a tropical tangle. In the San Juans, Oregon manroot is locally common on San Juan and Orcas Islands, growing on roadsides, scrubby areas, and open meadows to as high as 1700 feet on Mt. Constitution.

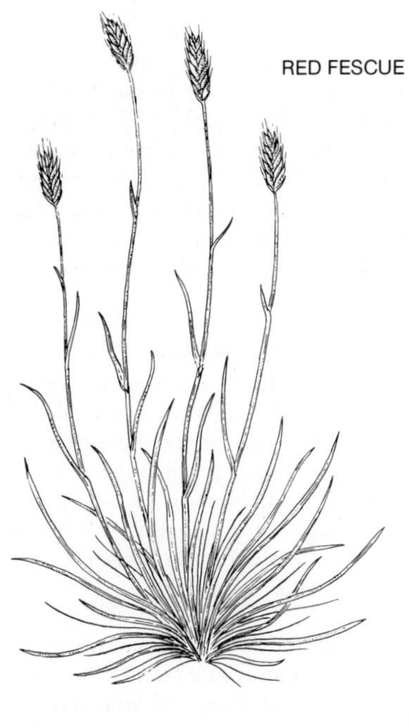

RED FESCUE

SMALL-FLOWERED PRAIRIE STAR

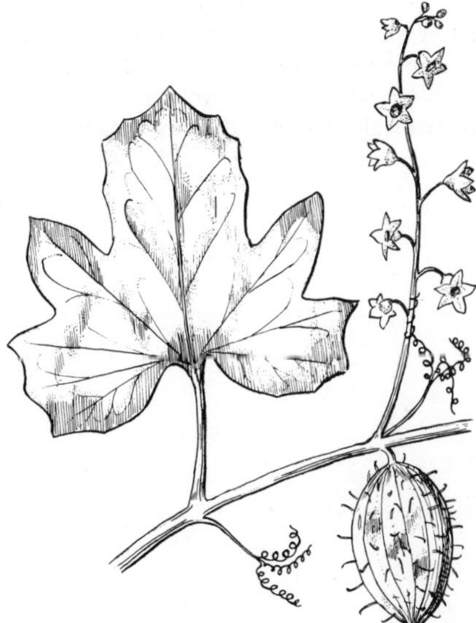

OREGON MANROOT

White (continued)

YARROW *(Achillea millefolium)*. Few plants have a greater range and vegetative variability than this one, occurring from sea level to alpine across the Northern Hemisphere. Also variable is the response to the plant's odor—some persons finding it pleasant and others not. Whatever the response, few will have trouble identifying Yarrow, as the flat, white-topped clusters, 2–4 inches wide, are a familiar, unmistakable constituent of sandy beaches, dry meadows, and roadsides throughout the area. The clusters, comprised of many papery little blooms, are often present throughout the growing season. Along the 1–3-foot stems are white-hairy, deeply dissected leaves, so soft that they seem featherlike.

HYACINTH BRODIAEA *(Brodiaea hyacinthina)*. Gracing a meadow with its snowy white bells, this is an enchantress of stunning loveliness. Finding one of the solitary, 2–3-foot beauties is achieved most easily during June and July. Common, especially on San Juan Island, where it is numerous enough to have spread to roadside ditches in a few places; its charms have not been lost on some local residents, who have taken to cultivating it, for which we must commend them. Howell's brodiaea *(B. howellii)* is equally attractive. It differs in having large, deep blue trumpet flowers. Watch for solitary plants in open meadows of San Juan Island, from Cattle Point north to Mt. Dallas–Cady Mt., and at a few other places; scarce on Orcas Island.

POISON-CAMAS *(Zigadenus venenosus* var. *venenosus)*. Steady companion of camas and fritillary, this wildflower is their match in beauty but contrasts sharply in that it has a poisonous bulb and foliage. Watch for the tightly packed raceme of small white flowers, present from April to June. When not blooming, note the long, grasslike leaves, similar to those of Great camas, but each with a deep central groove. The black-coated bulbs are a good signal that they should definitely not be eaten. To 24 inches tall.

GREENE'S REIN-ORCHID *(Habenaria greenei)*. Dense clusters of gleaming white flowers crowd the spike with such a dazzling grandeur that even the most indifferent person cannot help but admire them. Exhibiting considerable habitat variability, gorgeous plants will be found posing for photographers and artists on open, dry coniferous forest borders, in meadows, and occasionally on sandy beaches, but in all cases they choose hotter, more exposed sites than other rein-orchids. Leaves are absent except for the thick, scalelike ones hugging the 8–16-inch stems. Blooms from June until August. Common in the San Juans, growing on all larger islands and on many of the smaller ones.

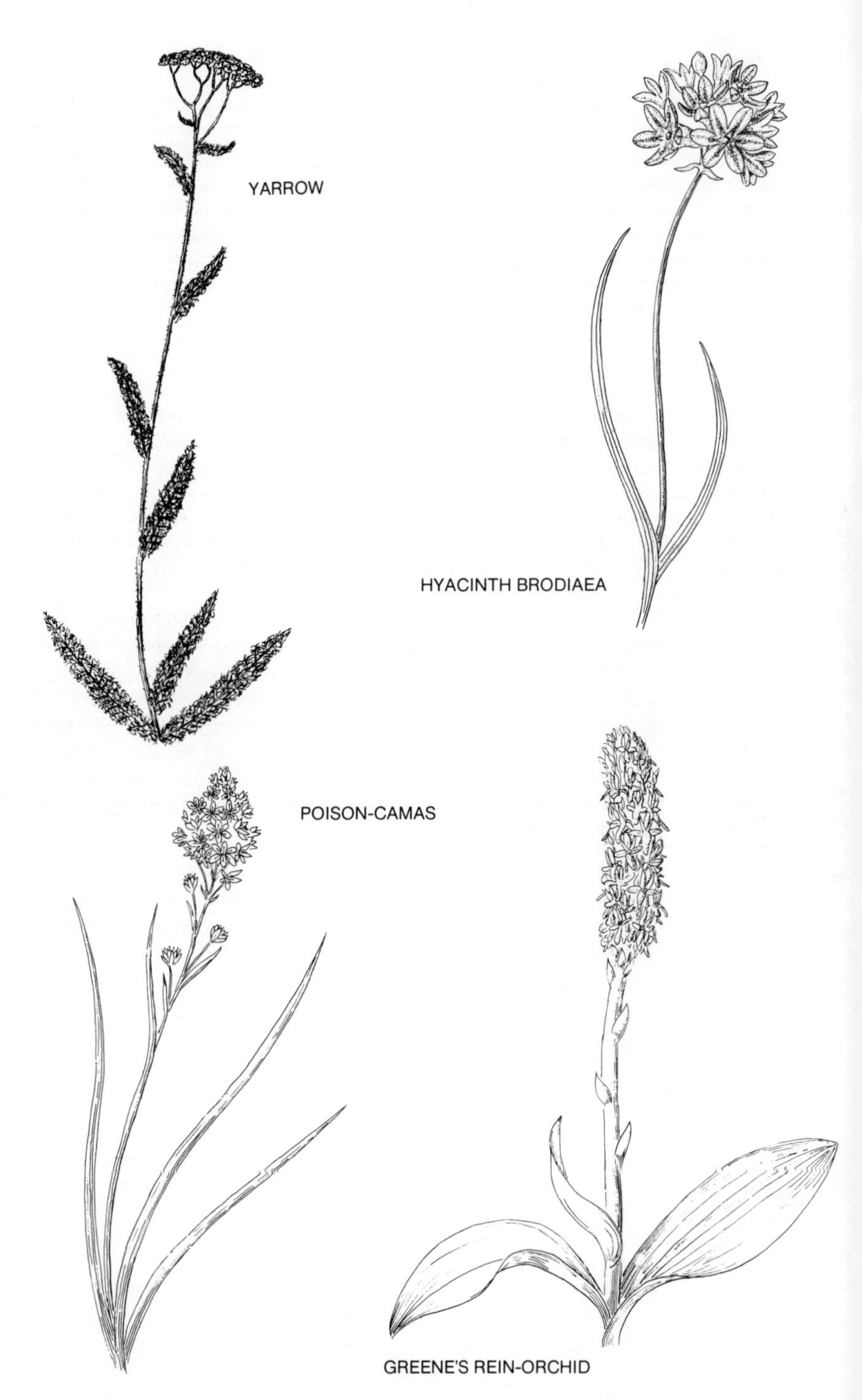
YARROW
HYACINTH BRODIAEA
POISON-CAMAS
GREENE'S REIN-ORCHID

Yellow

WESTERN BUTTERCUP *(Ranunculus occidentalis* var. *occidentalis).* A dense, brilliant yellow color covers many San Juan meadows in spring, thanks to the omnipresence of this species. The shiny flowers sit atop 6–24-inch stems, while the leaves vary from broad and trilobed at the base to narrow and dissected on the hairy stem. California buttercup *(Ranunculus californicus)* has eight–fifteen narrower petals, versus five–six ovate ones with *R. occidentalis,* and tends to be shorter. Found in meadows mostly from Cattle Point to Mt. Dallas on San Juan Island, and Iceberg Point to Davis Head on Lopez. Often grows, and sometimes hybridizes, with Western buttercup.

PRICKLY PEAR CACTUS *(Opuntia fragilis).* Although it might seem difficult to imagine cactus occurring naturally anywhere in Puget Sound country, it actually does grow at many places in the San Juans. Unwary travellers may become acquainted with it in a painful manner because the dull greenish or brownish clumps are well camouflaged and are easily sat or stepped upon; the sharp spines will inflict a memorable impression. However, this cactus sometimes comes out of hiding for a flamboyant announcement of summer's onset during June and July, when the tissue-papery, magnificent yellow flowers adorn steep rocky slopes. This is a sturdy little plant, seldom over 6 inches tall, which protects itself from being devoured by means of its spines. More often associated with the dry interior places east of the Cascades, this cactus' presence in the San Juans is true testimony to the dryness of the region.

NAKED DESERT PARSLEY *(Lomatium nudicaule).* Meadows and occasionally grassy beaches are home to this common San Juan herb. Coming upon the lemon-yellow blooms in early or mid-April is one of the more pleasing experiences one could have, although eating the thick, tasty, grayish green leaflets is more satisfying. Indians had an acute interest in this species as a staple food source, reflected in another name, Indian celery. Two other names, Cous, the Indian name, and Indian consumption plant, a reference to its use in combating tuberculosis among indigenous peoples in the past, are also used. To 24 inches tall.

WESTERN BUTTERCUP

PRICKLY PEAR CACTUS

NAKED DESERT PARSLEY

Yellow (continued)

SPRING GOLD *(Lomatium utriculatum)*. A vivid, canary-yellow is evident in the blooms of this well-named species. Justly celebrated, carpets of Spring gold decorate many a San Juan meadow and rock outcrop during April and May, sometimes lingering later. Attractively complementing the blooms are dark-green, shiny, fernlike basal leaves. Sometimes 12 inches tall. Iceberg Point, Lopez Island, is blessed with vast colonies of this parsley.

MENZIES' TARWEED *(Amsinckia menziesii)*. A rugged, spiny, tough plant whose image is forever etched in our minds as it grew along some dry, hard-packed roadside on San Juan Island. Small, yellow tubes emerge from a dense, curling spike, each bloom ¼–½ inch long and sometimes flecked with orange. Blooms from May to July; most common on San Juan and Lopez Islands. Rancher's tarweed *(A. intermedia)* is very similar, but the tubes are slightly larger, wider, more protruding, and are orange-yellow to orange. Generally prefers more upland sites. Both species may reach 3 feet in height.

Red

SHEEP SORREL *(Rumex acetosella)*. Although the loose racemes of tiny yellow and reddish flowers are attractive among dewladen grasses early on a spring morning, almost any habitat is suitable for the growth of this abundant European weed. Blooming through much of the season, Sheep sorrel is in fact daintier than other members of its genus, the docks, although rock gardeners may choose to overlook that feature after having weeded it out for the thousandth time. Note the distinctive arrowhead shape of the leaf bases. Although the leafage is edible, it is rather sour, due to the oxalic acid content. To 20 inches tall.

Pink

CRANE'S BILL *(Erodium cicutarium)*. The little rosettes of fernlike leaves and reddish stems that cover our rocky outcrops and meadows so abundantly belong to Crane's bill, so familiar now that it is difficult to imagine not seeing it on any spring outing. However, don't be fooled: this species is a Eurasian adventive, albeit an attractive one. During April through June, the little magenta flower stems rise 6–12 inches up, among the first of the spring pioneers to usher in the new season. Later the carpels (seed cases) form, ultimately flinging away the seeds by splitting in upward-curling strips. Both the genus and common names denote the carpel's similarity to a heron or crane's beak.

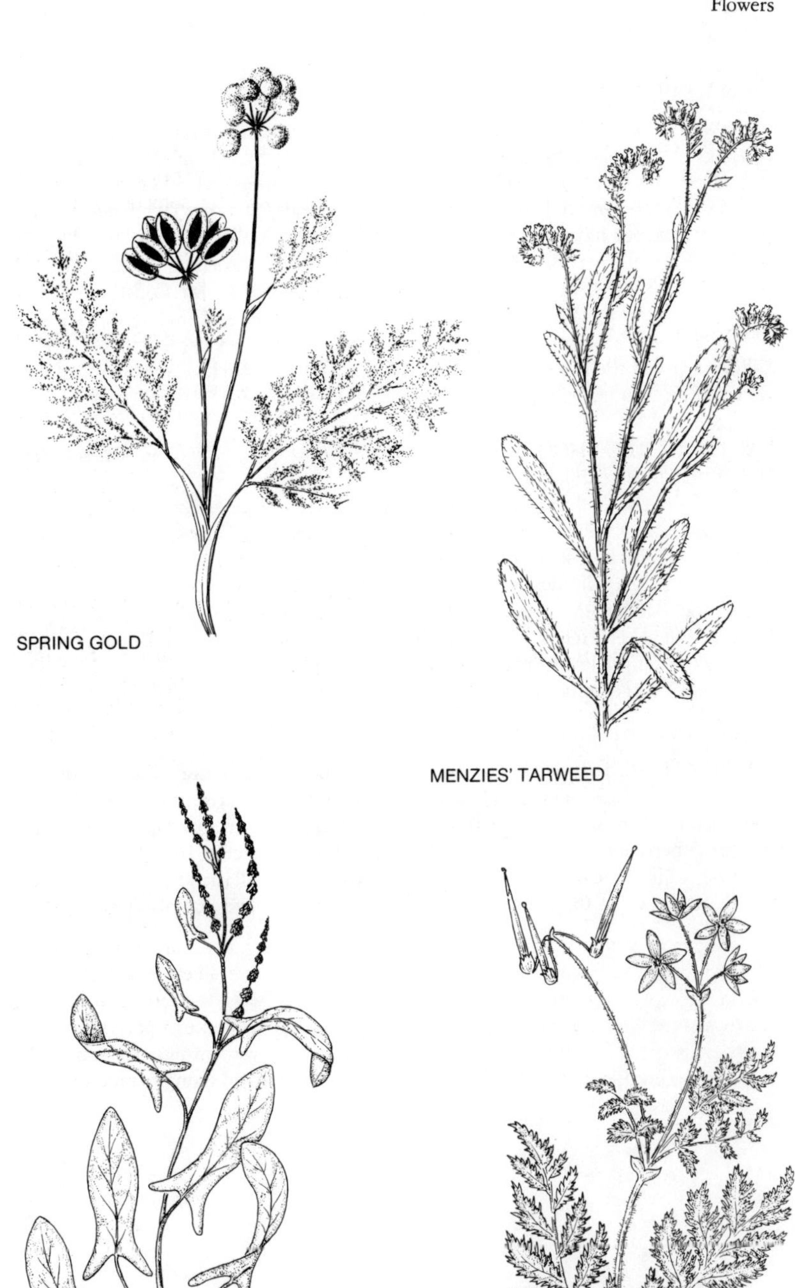
SPRING GOLD
MENZIES' TARWEED
SHEEP SORREL
CRANE'S BILL

Pink (continued)

FAREWELL-TO-SPRING *(Clarkia amoena).* The transition from spring's fresh, dewy meadows to the heated summer slopes is marked by the passage of this wildflower, a jewel of rare elegance. *Amoena* is the Latin word for charming, and indeed there are few epithets that seem more fitting for the pinkish lavender blooms. The petals of each cup-shaped flower have central reddish markings, a distinctive characteristic. Highly variable, the stems may be a few inches to 3 feet high and ascending or erect; flowers vary from ½–3 inches or more wide. The watchful observer on Mt. Constitution, San Juan Island, or in the Sucia Island group has the best chance of finding this species, especially if he visits a pristine meadow during June. The genus name honors Captain William Clark of the Lewis and Clark expedition, 1804-06. We have two forms of this species in the San Juans; var. *caurina* is widespread, var. *lindleyi* rather scarce.

FEW-FLOWERED SHOOTING STAR *(Dodecatheon pulchellum* var. *pulchellum).* One of the most exhilarating wildflower experiences the San Juans offer is the opportunity to take in the unbelievable blooms of this masterpiece. Each nodding flower is reminiscent of a falling star, the pinkish purple petals, like flames, trailing the dark point of stamens and style. A striking gold ring, highlighted by soft white above, completes the splendid design. Narrow to obovate leaves in a fleshy basal rosette precede the March-to-May spectacle. Grows in clusters on moist meadows, usually just above the salt water. Henderson's shooting star *(D. hendersonii)* is similar but perhaps even more beautiful. Its blooms are a sharp, flaming pink and borne above much wider ovate leaves. Most common on San Juan Island, where it prefers upland meadows. To 16 inches tall.

SEA BLUSH *(Plectritis congesta).* Sea blush, flushed in light rose-pink or glorious reddish purple, covers many San Juan hillsides with its varied accents from April until June. Each head consists of many tiny tubes; their tight arrangement is recognized in both the generic and specific names. The angled stems may reach 16 inches in height. Sea blush, while quite common on seaside meadows and small islands, does not always grow near the sea, as the name suggests; it is sometimes found on upland meadows.

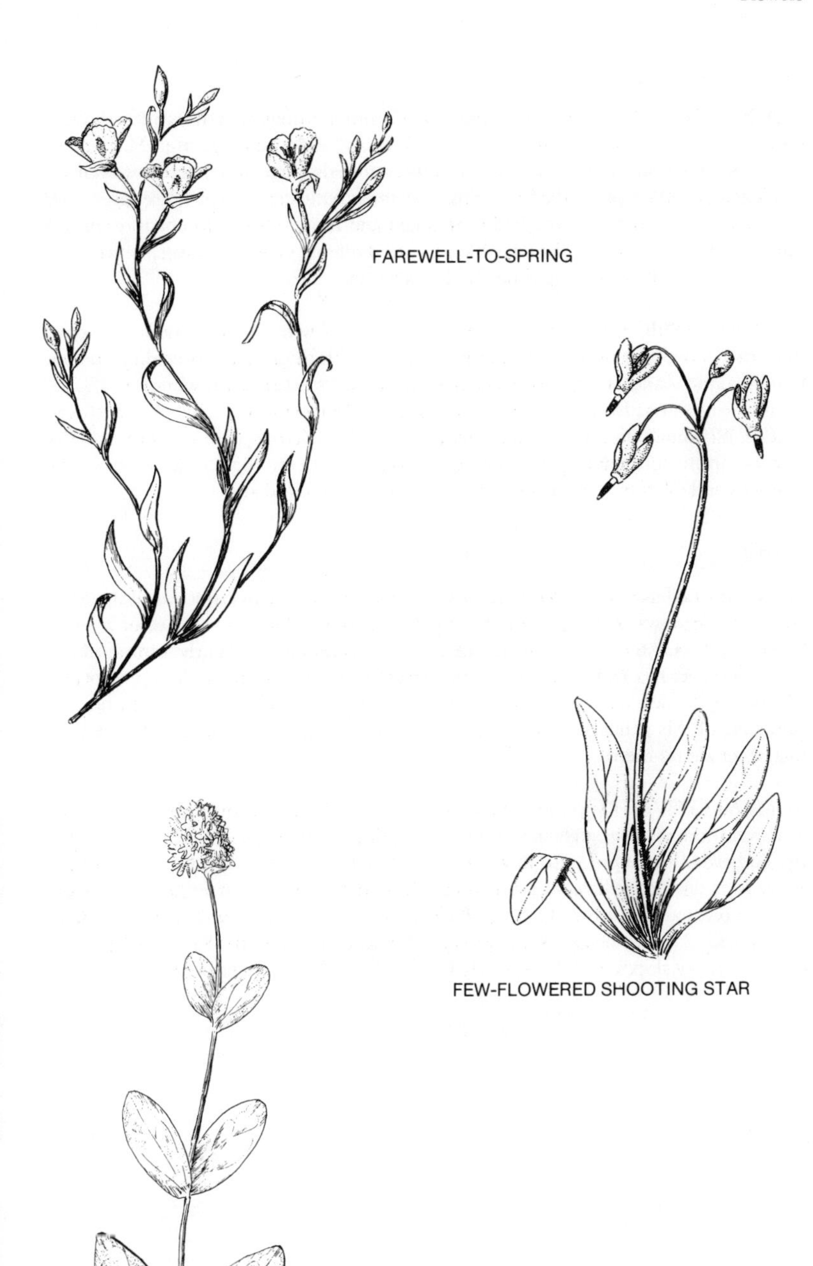
FAREWELL-TO-SPRING
FEW-FLOWERED SHOOTING STAR
SEA BLUSH

Pink (continued)

HOOKER'S ONION *(Allium acuminatum)*. Common throughout the islands, the rosy purplish blooms are a pleasant sight on meadows and rocky outcrops from May to July. Close examination of the urn-shaped flowers reveals three inner tepals (modified petals) with ruffled or toothed margins, and three outer tepals which are larger and recurved. Although edible, the bulbs are small and more difficult to obtain than one might think. Anticipating the May blooms are the long, grasslike leaves, which have usually withered by blooming time. To 12 inches tall.

NODDING ONION *(Allium cernuum)*. June and July witness the refined elegance of this onion's nodding pink blooms, often found in with those of the preceding species. Fragile stamens and the style protrude from the pink (or occasionally white) cups. Long narrow leaves arise from the bases of stems (up to 18 inches in height). A widespread species, Nodding onion grows in meadows, on rocky outcrops, and sometimes on dry forest margins in the San Juans and occurs well up into the higher elevations of the Cascades and Olympics. An important food source of the Indians.

Purple

MENZIES' LARKSPUR *(Delphinium menziesii)*. A spectacular attraction of the April and May meadows is this alluring beauty, the deep, rich blue to violet color a joy to behold. Not every meadow has it, but the species is widespread and fairly common. Each gem of a flower has a spur and two rounded teeth extending from the two upper petals. The deeply divided, hairy leaves are few along the stems, which are up to 16 inches tall. Members of this genus are poisonous to cattle, but sheep are immune and have been used to eradicate them.

TOMCAT CLOVER *(Trifolium tridentatum)*. One of the most refined and showy of the 15 clover species in the San Juans, Tomcat clover displays pleasing reddish purple, white-tipped heads from April until July. Ten–sixty tiny blooms make up each of the heads. Aside from distinctive flowers, two excellent field marks are the sharp-tipped calyx lobes and narrow, toothed leaves. Occasionally the creeping or ascending stems of healthier plants reach 2 feet in height. Widespread and quite common in the San Juans, growing on open, rocky slopes, meadows, sandy beaches, and even at times on weedy sites.

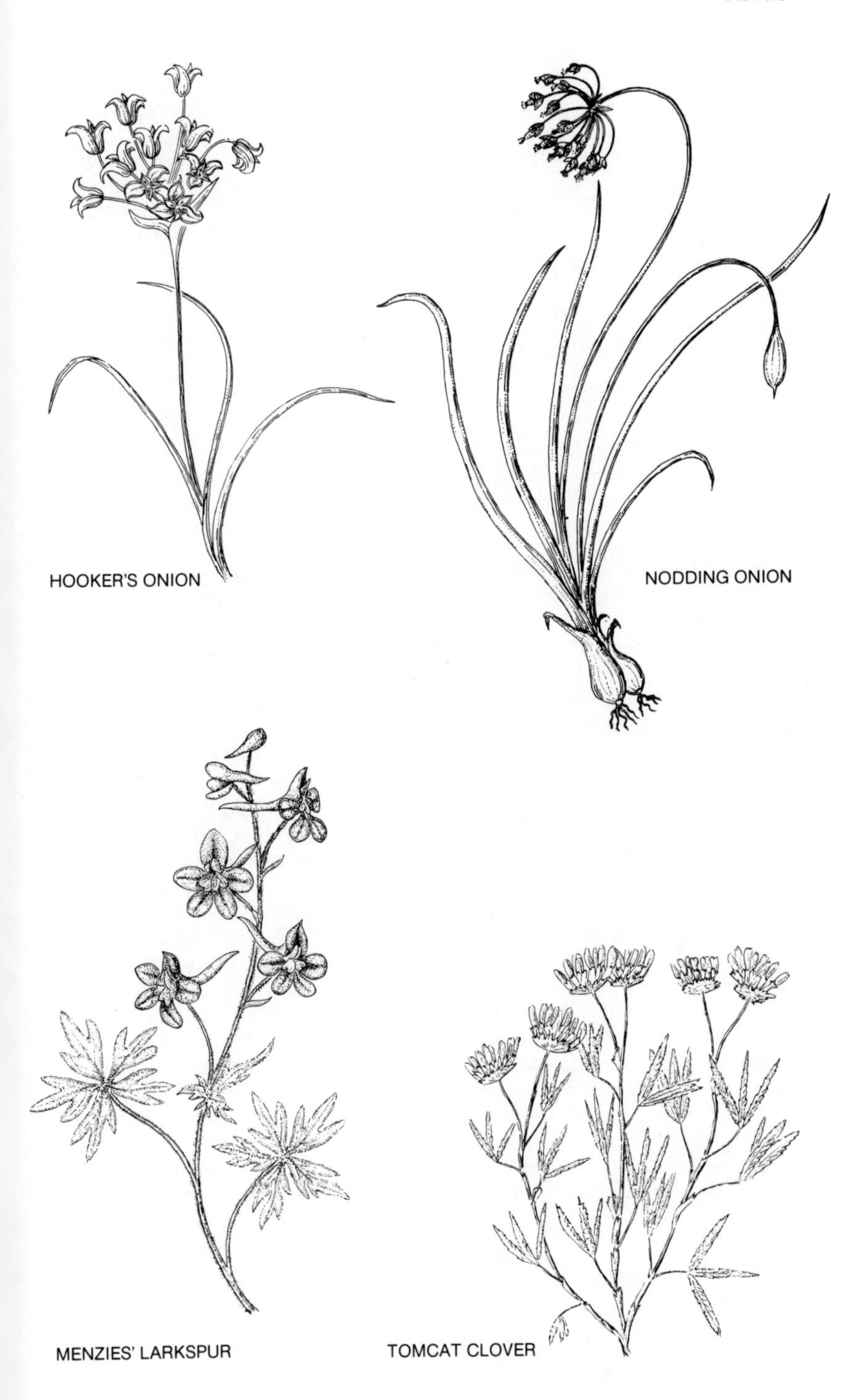
HOOKER'S ONION
NODDING ONION
MENZIES' LARKSPUR
TOMCAT CLOVER

Purple (continued)

WESTERN LONG-SPURRED VIOLET *(Viola adunca* var. *adunca).* The agreeable balance and full, deliciously purple flowers of this meadow dweller compete with those of Menzies' larkspur for the championship of the spring purples. At the Cattle Point meadows, this violet is almost abundant, for which we should be amazed, considering its competitors—the Canada thistle and a horde of aggressive grasses. In fact, this species is the most common of the San Juan violets, occurring on many meadows. Note the long, hooked spur that extends past the stem, a feature that distinguishes it from Howell's violet *(V. howellii),* a similar species. Howell's violet has a fatter spur that does not extend past the stem, or barely so; the leaves also tend to be more heartshaped, although this is a more variable characteristic. It is also slightly taller and larger flowered. Less common; prefers moist to dry coniferous forest edge and, less frequently, open meadows. To 8 inches tall.

HARVEST BRODIAEA *(Brodiaea coronaria).* Giving a lesson in felicitous simplicity is this lily, the waxy blooms pointing skyward from the tips of creeping stems during May to July. Each of the 1–3-inch rosy purple to violet flowers has an irrefutable attractiveness that cannot be overlooked in certain meadows. As with many lilies, each tepal (modified petal) has a purplish midvein. Also note the three white staminodia that encircle the three central stamens. To 10 inches tall. Both common and genus names honor James Brodie, Scottish botanist of the late eighteenth–early nineteenth centuries.

GREAT CAMAS *(Camassia leichtlinii* var. *suksdorfii).* Reigning beauty of many San Juan meadows, this wildflower dominates many localities in spring, and its blooms are well-known to islanders. Deep bluish or purple, the six-point stars blanket vast expanses in dense colonies, forming luscious contrasts with the canary-yellow blooms of the abundant Western buttercup. Clusters of 10–30 flowers will be found on each 1–2-foot stem. A staple food source of many Northwest Indian tribes, the camas root was so valued that wars were fought over the best grounds and, periodically, fires were set to beat back encroaching woodlands and thus ensure future harvests. The latter practice is believed to be at least partially responsible for maintaining the numerous San Juan meadows. Indeed the large bulbs are excellent eating, with the taste of a creamy potato; however, we discourage the observer from consuming them for several reasons. First, digging them up is often quite difficult, and several colonies have been decimated by over-eager food gatherers. Also, although Great camas' bluish to purple blooms are diagnostic, its bulbs resemble those of the white Poison-camas *(Zigadenus venenosus),* so that after blooming there might be confusion. Occasionally one finds stunning white forms of Great camas. The Latin name honors two Germans, Max Leichtlin, a German horticulturist of the nineteenth century, and Wilhelm Suksdorf, an early Northwest botanist who collected in Skamania County, Washington, particularly. Purple camas *(C. quamash)* is often confused with Great camas. It has slightly irregular flowers, as the lowest tepal curves away from the stem and is longer than the others. The separate withering of the tepals after blooming, unlike those of *C. leichtlinii,* which twist together around the ovary, is probably the best field mark. Less common in the same habitat.

WESTERN LONG-SPURRED VIOLET

HARVEST BRODIAEA

GREAT CAMAS

Purple (continued)

BLUE-EYED GRASS *(Sisyrinchium angustifolium).* Poking out of a seam on the sheathing stem, the bluish purple to violet "eye" (petals), complete with a yellow "pupil" (center), charmingly looks up at the passerby. It should be apparent to all that this is not a grass at all (and, in fact, is a member of the Iris family). Usually found in soggy pockets of the upland meadows during April and May. The stiff, flattened leaves and well-defined tips of the petals are good field marks. To 18 inches tall.

Brown

DWARF OWL-CLOVER *(Orthocarpus pusillus).* An observer intent on the showy meadow bloomers will easily overlook this little plant, since it is only 2–7 inches high. Despite the seeming vulnerability, it somehow manages among its larger neighbors and, in fact, does so with a vengeance, forming dense colonies at many localities. Lately Dwarf owl-clover seems to have increased to the point where one will even find it on weedy lawns, where it is the brave companion of plantains, Hairy cat's ear, and other "friendly" weeds. Blooming from April to June, Dwarf owl-clover has minute chestnut flowers, which lie nestled among the bractlike, hairy, purplish leaves.

CHOCOLATE LILY *(Fritillaria lanceolata).* Chocolate well describes the color of this distinctive member of the meadow community, although the flowers' dark brown mottling is splashed with greenish yellow and purple. The curiously bowl-shaped flowers (1–2 inches wide) are scattered or solitary on the 1–2-foot stems like so many little umbrellas. Note also the whorled lanceolate leaves (hence the specific name). Blooming during April and May, this lily is common throughout the archipelago. Iceberg Point, Lopez Island, is blessed by an especially large colony that features occasional banana-yellow forms. The bulbs are edible and covered with small, ricelike white nodules, the origin of another name, Rice-root. Kamchatka fritillary *(F. camschatcensis)* is larger and has darker, unmottled tepals (modified petals), with ridges along the veins of the inner surface. Perhaps now extirpated in the San Juans, it should be watched for. Historic records are from Orcas and San Juan Islands.

BLUE-EYED GRASS

DWARF OWL-CLOVER

CHOCOLATE LILY

2
OPEN ROCKY OUTCROPS

Few visitors to the San Juans fail to notice the omnipresence of rocky outcrops which, along with the dry climate, give the islands a distinctive character unknown elsewhere in western Washington. Nature's rockform artistry is evident everywhere, with rounded rocky knolls, steep, stony slopes, and gravelly prairies present. The rocky masses create a sense of time and permanence, as they yield only very gradually to the wave, the root, and other erosional pressures. Floral communities, developing slowly on the rocky outcrop, reflect the difficulty of colonizing such a habitat. Particularly on the south-facing outcrops, subject to less favorable growing conditions (for many trees and shrubs), the flora is predominantly low and herbaceous, bearing close resemblance to, and intergrading with, that of the south-facing meadow.

Glaciers sheared many of the sharp ridges and irregularities off the San Juan rock masses, leaving behind rounded, shalelike benches and outcrops. Small deposits, or outwash, were left in rock crevices and depressions, and later these sites became the focal point for vegetative development. Lichens and mosses coated the bare rock and soil pockets during the earliest phase, and many sites have progressed little beyond this stage today. Joining the lichens and mosses are various grasses and herbs, among which are some of the most appealing San Juan species. Decaying plant matter deepens and enriches the soil pockets, and when combined with the subsequent erosion along weakness planes in the rock masses, branching soil networks begin to crisscross slopes. If enough soil accumulates, shrubs and, finally, trees may appear. However, the timeless progression of lower (moss, lichen) to higher (herb, shrub, tree) vegetative development has several qualifications. First, the steepest, most exposed rocky outcrops were never colonized by shrubs and trees, for a number of reasons. Erosion—from rain and wind particularly—has kept soils too thin for trees and shrubs to grow, especially at steep or exposed sites where soils are more easily dislodged. Second, many sites are too sunbaked in summer and too low in nutrients and moisture levels. Third, forest encroachment on the open outcrops was kept in check by periodic fires, and still is by sea and wind-inflicted damage. Indeed, a windstorm occasionally topples trees as if so many dominoes, revealing thin soil and bedrock below. Finally, man-made alterations, including farming, tree cutting, etc., prevent forest encroachment as well.

Among trees, Garry oak *(Quercus garryana),* Rocky Mountain juniper *(Juniperus scopulorum),* and Pacific madrona *(Arbutus menziesii)* are most typical of exposed rocky outcrops. All three anchor well in thin soils and prefer sunny, well-drained sites, although Pacific madrona is more variable in these respects. Among herbs, many spring ephemerals are present because winter rains leave the soil beds moist in spring. As the summer drought arrives, sites become dehydrated, and the dynamic floral displays wither away. Only a handful of early summer bloomers, capitalizing on trapped moisture deep in the rocks, remain. However, in addition to the spring ephemerals, described largely under the south-facing meadow section, there are other interesting species which are more specific to rocky soils.

Wallace's Selaginella *(Selaginella wallacei)* is one such species, an attractive creeper that covers many rocky lichen balds. Gold-back fern *(Pityrogramma triangularis),* a distinctive, black-stemmed denizen of rock crevices, is a noteworthy member of the rock-garden community. Licorice fern *(Polypodium glycyrrhiza)* is most common on shady, moss-covered slopes of northerly orientation, while Field chickweed *(Cerastium arvense)* forms climbing mats on sunny slopes. Broad-leaved stonecrop *(Sedum spathulifolium)* and Small-flowered alumroot *(Heuchera micrantha)* are inevitably found on virtually any substantial rocky site. Two interesting parasitic species are the California and Naked broomrapes *(Orobanche californica* and *uniflora,* respectively), the first predominantly attacking Puget Sound gumweed *(Grindelia integrifolia),* and the other parasitizing stonecrop. Many other species, of course, are to be expected.

Rocky outcrop vegetation immediately adjacent to salt water differs somewhat from that of the upland areas. Intense wind and salt spray create additional stress on the habitat, but Sea plantain *(Plantago maritima),* Lance-leaved stonecrop *(Sedum lanceolatum),* and Sea thrift *(Armeria maritima)* are to be expected.

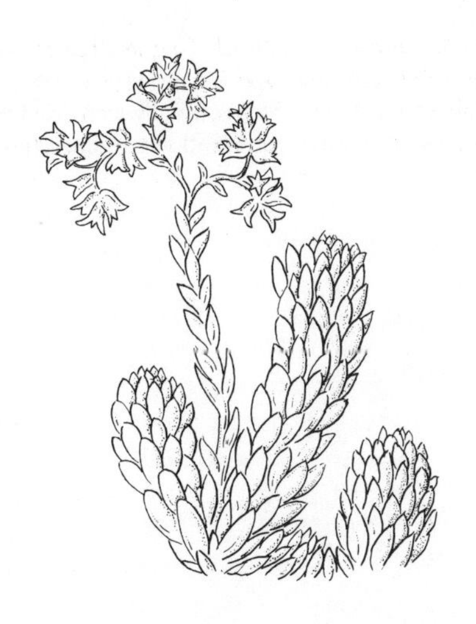

FERNS AND FERN-ALLIES

WALLACE'S SELAGINELLA *(Selaginella wallacei)*. A common species throughout the archipelago on exposed to fairly shaded sites, often forming a dense rock cover with various lichens and mosses. The evergreen stems are rigid, loosely branching, and freely rooting; they tend to form open mosslike mats. The short branches are equipped with small, stemless, linear leaves. Thus far this attractive rock climber is the only *Selaginella* known for the islands.

GOLD-BACK FERN *(Pityrogramma triangularis)*. An eye-catching fern, the tufted, blackish lustre of the stems reminding one of so many lengths of shiny wire as they rise from the rock crevices. Arising from short rhizomes to a height of 16 inches or so, the stems support distinctively triangular fronds. The leaflets are dark green above and yellowish-powdery below; the late summer drying and curling up of the fronds to reveal this yellowish underside has earned this fern its common name. Occurs on all the larger islands and many of the smaller ones.

LICORICE FERN *(Polypodium glycyrrhiza)*. Although the rhizomes are edible and were eaten by certain Northwest Indians, the licorice taste is too acrid for most persons. A rather nondescript species: the fronds grow to 12 inches long and the leaflets to 2–4 inches or so. Due to dryness and other factors, this fern is peculiarly absent as an epiphyte on moss-laden Bigleaf maple *(Acer macrophyllum)* in the San Juans, whereas in most of western Washington it is very common as an epiphyte on that tree. However, Licorice fern is common throughout the islands on heavily mossed-over rock ledges and sometimes on mossy forest margins. The growth cycle is a bit unusual: new green fiddleheads uncurl with the first fall rains, persisting through winter, and then withering with the onset of warmer weather during May–June. Mountain licorice fern *(P. amorphum)* also occurs widely in the San Juans, though less commonly. The fronds are shorter, with rounded, thicker leaflets. Watch for it in similar habitats, though it seems to prefer steeper, more open rock faces, where it grows in tight crevices.

GRASSES AND GRASSLIKE PLANTS

LITTLE HAIRGRASS *(Aira praecox)*. A dwarf of a grass that grows abundantly on the driest, hottest rocky sites, where few other species can survive. The panicle seed cluster is compact, spikelike, and about 1 inch long. Stems, usually seen dried and crisp, rarely pass 6 inches in height. European; also grows on meadows and beaches.

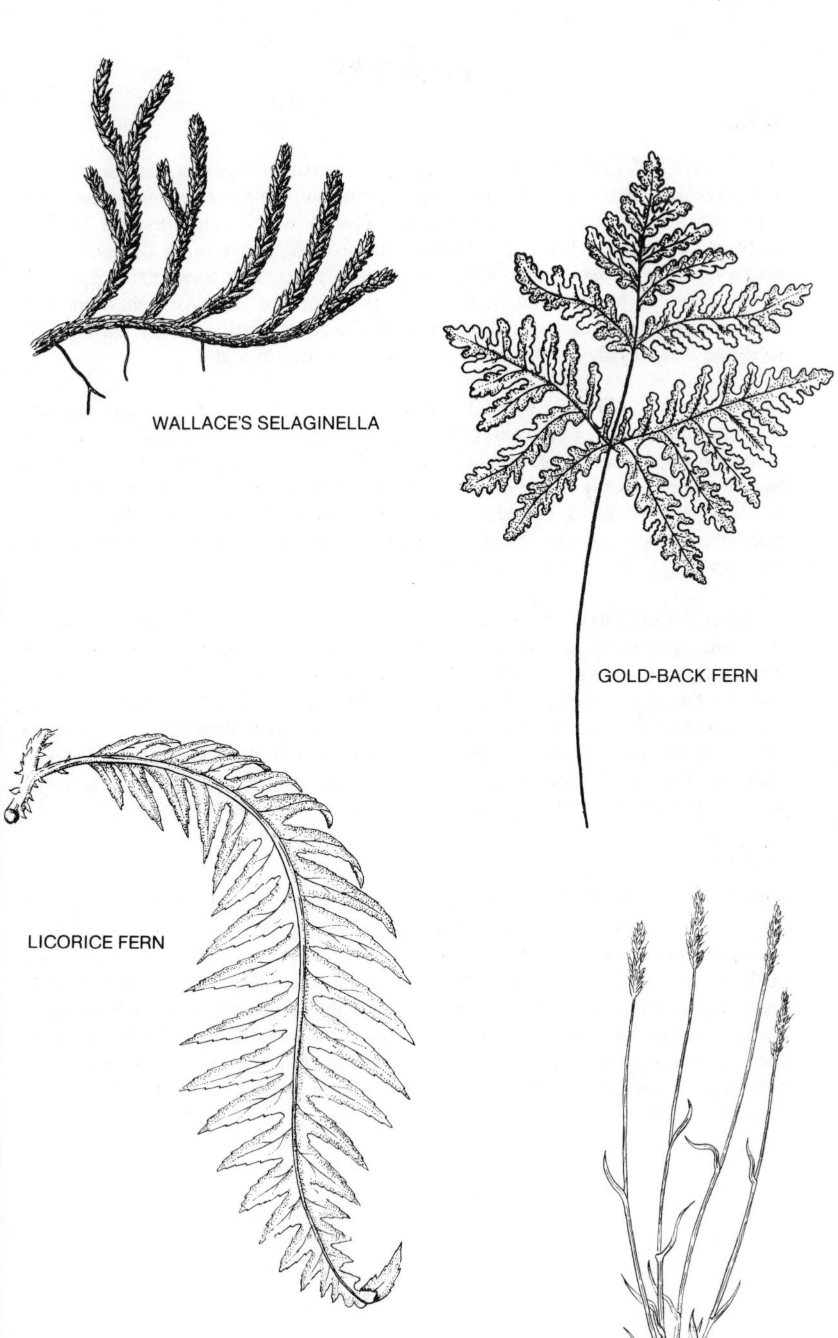
WALLACE'S SELAGINELLA
GOLD-BACK FERN
LICORICE FERN
LITTLE HAIRGRASS

FLOWERS

White

FIELD CHICKWEED *(Cerastium arvense).* An essential ingredient in the striking array of color evident on the San Juans' spring meadow, showy white is enthusiastically represented by the blooms of this wildflower. Each flower, ½ inch or so wide, has five deeply notched petals. Forming little mats, the creeping stems, often 12 inches long, readily climb up the rocky faces. They bear numerous fuzzy little leaves that are grayish to light green. By July, the little mats turn crispy and brown, a familiar sight of the hot, exposed outcrop. Found statewide, Field chickweed is common from sea level to subalpine and is easily champion of the genus *Cerastium* in beauty.

SMALL-FLOWERED ALUMROOT *(Heuchera micrantha* var. *diversifolia).* Coming upon the airy, cream-colored spray of little globes along some shaded, rocky slope is one of the more delightful wildflower experiences. From May to July, the open panicles of blooms will commonly be encountered, often with stonecrop. When out of bloom, the basal rosette of shallowly lobed hairy leaves is distinctive. Below the 1–3-foot reddish-hairy stem is a long taproot, which is reputed to taste of alum, hence the name. Found from sea level up to 7,000 feet in the Cascades.

PRAIRIE SAXIFRAGE *(Saxifraga integrifolia* var. *integrifolia).* The simple grace of this spring ephemeral is a most appreciated addition to our soggy, early spring meadows and moist, rocky slopes. Life is brief for the tiny white flowers that are borne in tight panicles: they appear in early or mid-April and have usually disappeared by June 1. The specific and variation names refer to the uncleft, simple leaves. Both the 6–12-inch stem and the leaves are very hairy. This species could possibly be confused with Western saxifrage *(S. occidentalis* var. *rufidula)* that has larger, toothed leaves and an open flower arrangement. Mostly confined to steep, shaded rocky slopes on Mt. Constitution.

Yellow

LANCE-LEAVED STONECROP *(Sedum lanceolatum* var. *nesioticum).* A regional endemic to the San Juans and adjacent British Columbia, var. *nesioticum* is widespread throughout the islands, though it is far less common than the following species. The key field marks of var. *nesioticum* are the generally smooth, purplish, overlapping (especially at the base) leaves which are blunt tipped, succulent, and round in cross-section; they are not brown and chaffy when dried. Also note the loose clusters of yellow flowers, which appear during early summer. Sometimes grows to 10 inches tall. Typically a denizen of rocky banks in close proximity to salt water, and somewhat more common on small islets.

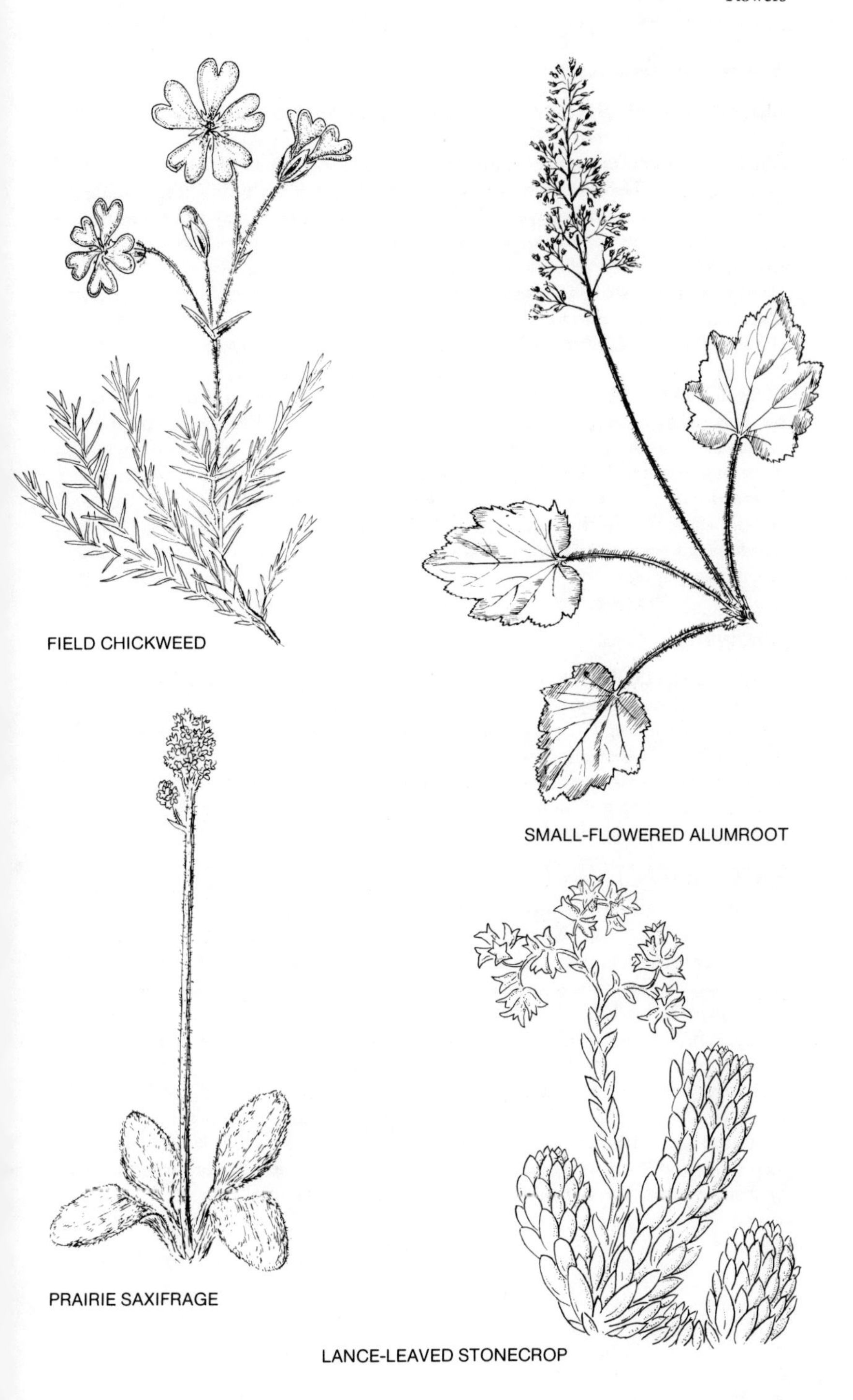
FIELD CHICKWEED
SMALL-FLOWERED ALUMROOT
PRAIRIE SAXIFRAGE
LANCE-LEAVED STONECROP

Yellow (continued)

BROAD-LEAVED STONECROP *(Sedum spathulifolium).* What species could be more well-known to the San Juan naturalist walking the rocky slopes than this one? This climber is in fact very common throughout the area and is not apt to be confused with anything else. The thick, succulent, evergreen leaves form little clumps along the creeping rootstocks and vary in color from a pleasant pearly gray to various hues of green, usually olive or lime; occasionally purplish plants will be found. The compact, flat-topped clusters of yellow blooms grace many a slope during May–June. Each of the "star" flowers has sharply pointed sepals and is seldom over 6 inches tall. An attractive species and an excellent addition to the rock garden; closely related to many well-known cultivated species of *Sedum* and also the famous "Ice plant" from Monterey and vicinity.

COMMON WOOLLY SUNFLOWER *(Eriophyllum lanatum* var. *lanatum).* A cheery spring wildflower bringing a sunny hue to many San Juan hillsides. Quite common on open rocky sites or gravelly outwash meadows, usually growing in showy clusters up to 12 inches high. The brilliant yellow flowers are very apparent from their open outcrop haunts during June and July. The leaves are entire to deeply lobed, grayish green, and densely white-woolly, the latter a good field mark and recognized in the genus name, which means woolly leaf. Fairly tolerant of disturbance, unlike many other rocky outcrop and meadow species.

Orange

CALIFORNIA POPPY *(Eschscholzia californica).* Large, brilliant, and shiny—perhaps no other words more aptly describe the orange-to-yellow flowers of this conspicuous, well-known species. The grayish green, finely divided leaves are less noticeable but linger well after the dazzling blooms have withered. Locally abundant at several San Juan localities, such as the west side of San Juan Island, where dense colonies paint certain hillsides in vivid orange during spring. Widely introduced in western Washington and thoroughly naturalized in many places.

HAIRY INDIAN-PAINTBRUSH *(Castilleja hispida* var. *hispida).* Displaying brilliant reddish to bright orange-tipped bracts, the memorable Hairy Indian-paintbrush has earned more than one wayfarer's admiration. Blooming from April to as late as September, this widespread paintbrush is most apt to be found on semi-open rocky slopes, open meadows, or occasionally forest margins, although it is not overwhelmingly common in the San Juans. Like many of the paintbrushes, this species is mildly parasitic on the roots of nearby grasses and herbs, surely part of the reason for its enduring blooms. Although only one form occurs in the San Juans, this is a variable wildflower, with several variations occurring well up into the Cascades, from British Columbia to northern Oregon. Variability is a characteristic of much of the genus and so is hybridization, and the two together make the constituent species of the group difficult to identify, as any Northwest botanist will tell you. However, only two or three species are present in Puget Sound country, and this one seems to be most common.

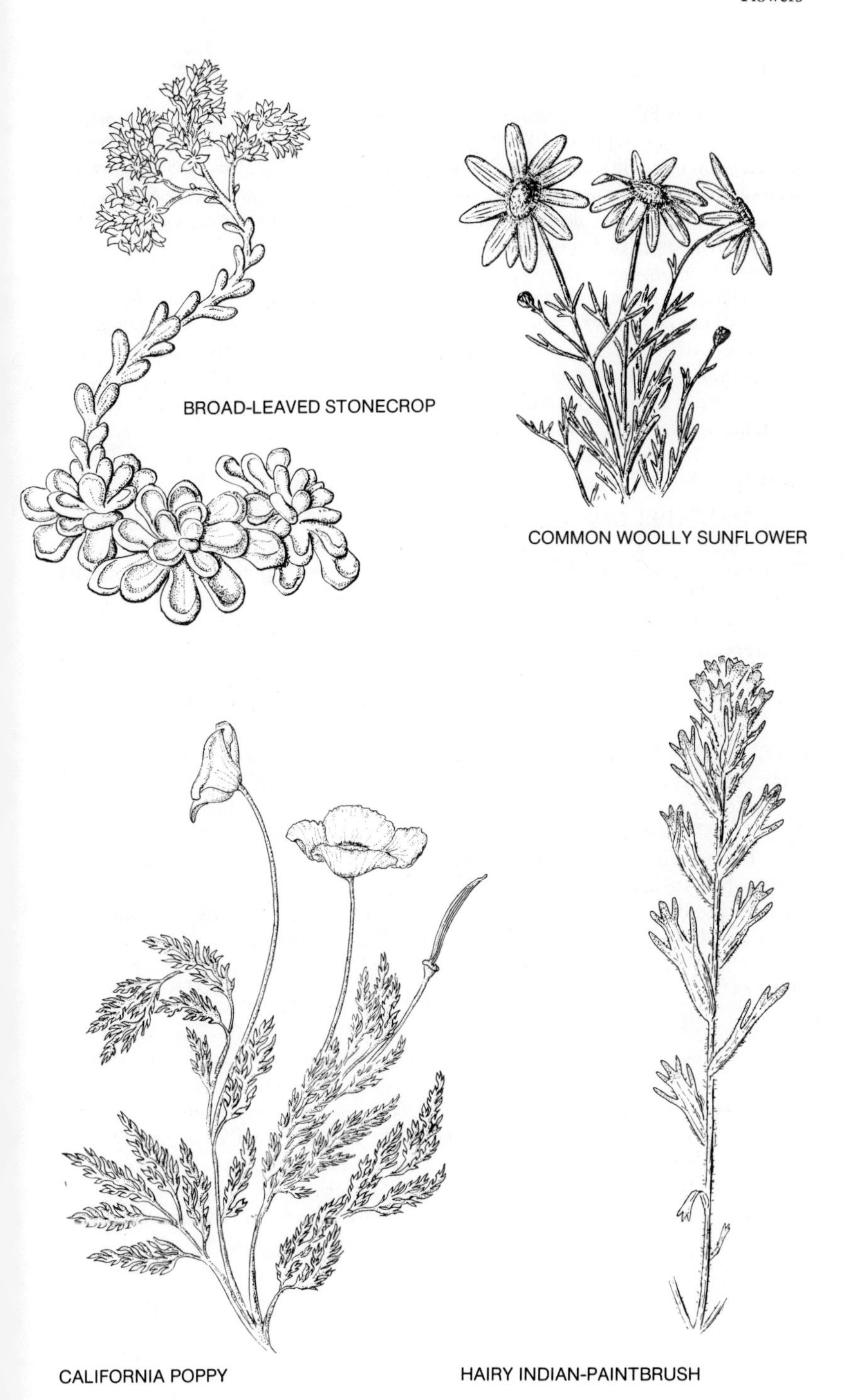

BROAD-LEAVED STONECROP

COMMON WOOLLY SUNFLOWER

CALIFORNIA POPPY

HAIRY INDIAN-PAINTBRUSH

Pink

FALL KNOTWEED *(Polygonum spergulariaeforme)*. Long after the colorful spring ephemerals have disappeared from the rocky outcrops, this meek little herb begins blooming, usually in July, and often does so well into October. The tiny pink bell flowers, in groups of one to four, arise on very short (1–6 inches) stems out of the leaf axils. The linear leaves are glossy, deep green, and leathery. Abundant throughout but easily overlooked among dried grasses.

SEA THRIFT *(Armeria maritima)*. A rugged, tufted small plant that likes the windy, rocky bluffs and meadows by the sea. From April to August the pinkish, compact heads, reminiscent of onion, and the basal clumps of linear, needlelike leaves are diagnostic. The common name points out the plant's thriftiness—its ability to grow in thin, gravelly soils exposed to strong winds and salt spray. Found over much of coastal North America and Eurasia; common in the San Juans.

Purple

CALIFORNIA BROOMRAPE *(Orobanche californica)*. Watch for the tight clusters of light pinkish purple flowers on rocky slopes just above the salt water from May to July. The tufted, stout stems reach 5 inches, provided they are well supplied with the nutrients and juices of Puget Sound gumweed *(Grindelia integrifolia)*, the preferred host of this parasite. California broomrape is a peculiar species, both in habits and appearance, and is worthy of being watched for; it is fairly common in the San Juans, showing a particular propensity for small islets and places like Davis Head, Lopez Island, or the Friday Harbor Laboratory grounds. Our most common form is var. *californica.*

NAKED BROOMRAPE *(Orobanche uniflora)*. An outstanding meadow and rocky outcrop dweller that owes the beauty of its rich, deep purple or rosy purplish flowers to the stonecrops, which it parasitizes. A frontal view reveals the alluring, rich yellow throat, an added attribute of a wildflower that is with us too briefly during April and May. The one–three pedicels, supporting one flower each, are easily mistaken for the stems, which are in fact much smaller and often scarcely emerge above the surface. This is a widespread San Juan species, and certain rocky meadows have dense colonies, as on Orcas Island. Seldom grows over 4 inches tall. Our common variety is var. *minima.* Occasionally attacks cultivated stonecrops, like Wall-pepper *(Sedum acre)*, as in Friday Harbor.

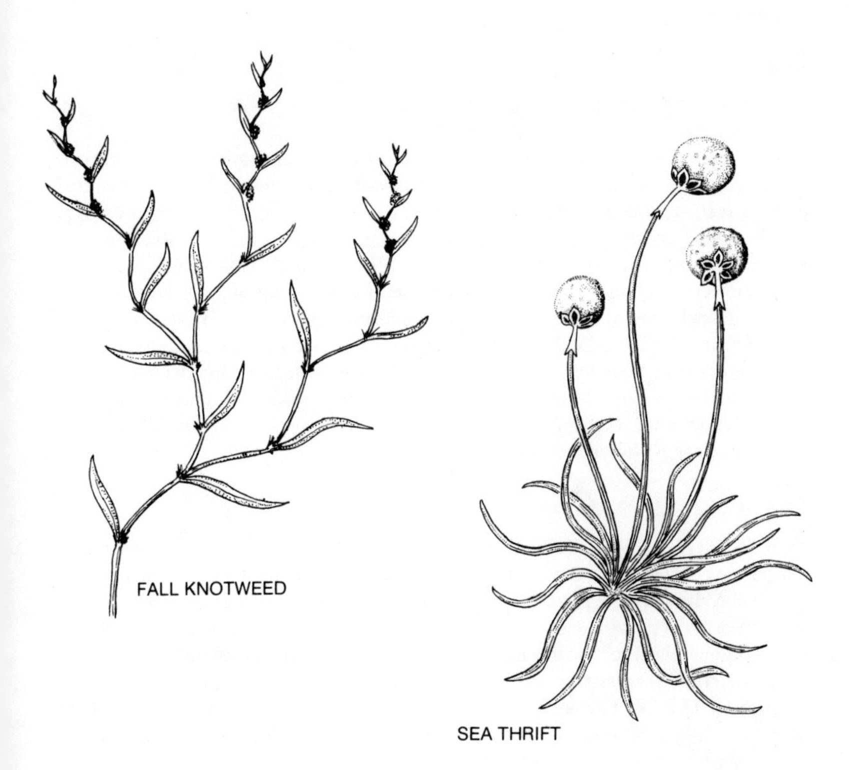

FALL KNOTWEED

SEA THRIFT

CALIFORNIA BROOMRAPE

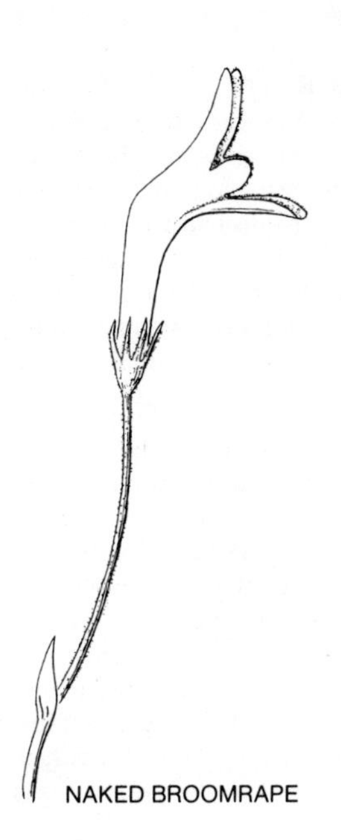

NAKED BROOMRAPE

Blue

BLUE-EYED MARY *(Collinsia parviflora)*. An early harbinger of spring's arrival is the little Blue-eyed Mary, also called Small-flowered collinsia. Often as early as March, clustered masses of the tiny flowers are found on open, moisture-laden rocky outcrops. Each bloom (¼–½-inch long) has lovely blue lower "lips"—the upper lips are whitish, giving a hint of a grayish or light bluish sheen. The horizontal, inflated flower tubes are attached at an oblique angle to the sepals. Flower arrangement is variable: clustered, solitary, terminal, or axillary forms can be found. Seldom grows to 8–10 inches in height. Large-flowered collinsia *(C. grandiflora)* is very similar but slightly larger, having flowers ½ inch long or slightly longer and sometimes reaching 16 inches in height. A good field mark is the attachment of the flower tubes, at a more pronounced right angle to the calyx. Also, the flowers are arranged in a showier, whorled fashion. Less common; occurs in similar habitats, although sometimes is found on open grassy bluffs. Both species bloom until May and share a genus name that honors Zacheus Collins (1764-1831), an early American botanist.

Green

NORTHERN WORMWOOD *(Artemisia campestris* ssp. *borealis* var. *scouleriana)*. The velvety-soft leaves are a warm, silvery gray color, displaying slight variations as they are touched by a breeze. A sturdy perennial with clustered stems that may reach 2 feet. As with many of the *Artemisia* group, a pleasant scent arises from the foliage. Erratic in distribution and habitat: as much a member of the maritime community as that of the rocky outcrop, showing up on beaches in many places; also grows on gravelly, moderately disturbed prairies. Locally abundant in the San Juans, forming dense colonies as it spreads by substantial rootstocks. A member of the same genus as Sagebrush.

TREES

ROCKY MOUNTAIN JUNIPER *(Juniperus scopulorum)*. A cedarlike tree restricted mostly to the San Juans in western Washington, where it occurs along dry, rocky bluffs adjacent to salt water. Often shrubby and gnarled, each tree frequently has two or more trunks, growing in an open, scraggy manner. A torn-off strip of the grayish brown bark reveals a rich, reddish brown inner layer, which has a fragrance strongly reminiscent of Western red-cedar. The scalelike needles are short and paired—they too suggest those of cedar. The pale blue, sweet-scented berries are readily eaten by birds, and seed dispersal via bird droppings is an essential component in the tree's reproduction. Certain humans have also been known to utilize the berries in making gin.

GARRY OAK *(Quercus garryana)*. What a handsome tree this is, especially when leathery, dark green leaves shine in the sunlight! This feature and the distinctively open, spreading growth form make it recognizable from some distance. Underneath the gray, fissured bark is a fine-grained wood that is very heavy and very strong; were it not for the general scarcity of the tree in Washington, it would probably be of significant commercial value here. Its distribution is erratic, with substantial stands scattered about at Sequim, near Tacoma, at Naches–Yakima, from Olympia south, and, of course, in the San Juans, where it is widespread on shallow, dry, rocky soils, associating with Pacific madrona *(Arbutus menziesii)*. The ovoid, 1-inch acorns are mass-produced, providing food for a remarkable entourage of insects, mammals, and birds. David Douglas, who discovered this tree in the 1820s, named it for Nicholas Garry, a Hudson's Bay Company official who was Douglas' personal friend and co-traveller.

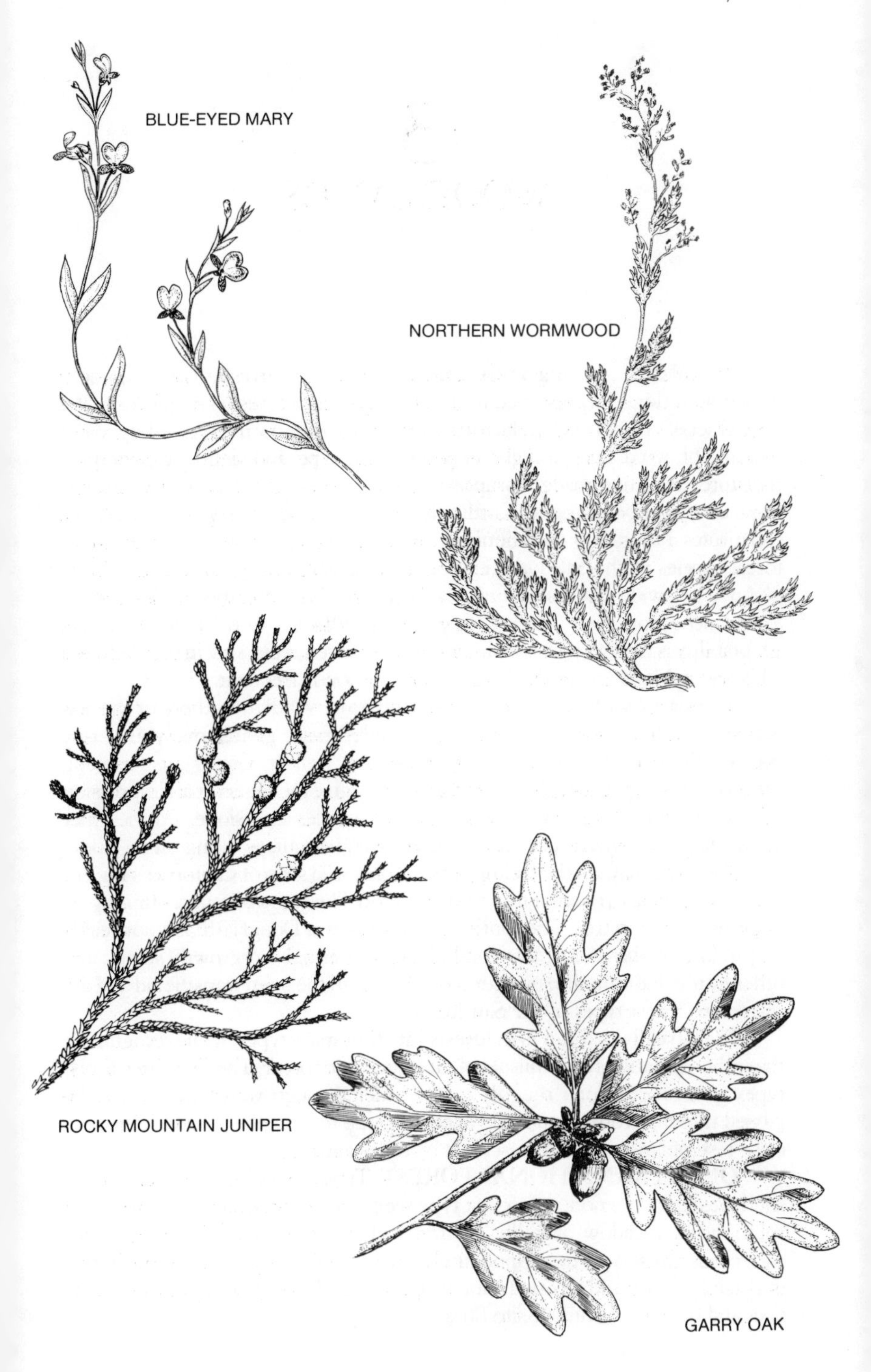
BLUE-EYED MARY
NORTHERN WORMWOOD
ROCKY MOUNTAIN JUNIPER
GARRY OAK

3
WOODLANDS

Woodlands, covering much of the land area, form an essential component of San Juan flora. Represented in several types, each forest is a reflection of a special set of environmental factors. Floral composition varies according to the amount of wind and sunlight exposure, soil type and depth, topography, moisture, and man-made disruption. As the forest environment is altered, floral composition changes accordingly; indeed, one might say that each forest constitutes a mosaic of communities changing through time and space. Many forest species in the San Juans exhibit considerable ecological adaptability by occupying a wide environmental gradient, and thus attempts to describe an exact niche for each species may prove difficult. Consistent with this adaptability is the fact that transitions between forest types and, in fact, between all island habitats are gradual, and intergradation is frequent.

Compared with other western Washington woodlands, those of the San Juans are relatively dry and lacking in underbrush, giving many an easily traversable, parklike character. Open canopy, well-lit forests are frequent, especially at steep, rocky sites. Such forests may give the observer an impression of having seen a forest more typical of the Cascades' east slope. On the other hand, in places where oak and juniper straddle valleys along steep, rocky coastlines, one may get the feeling of being on a wild strip of California's Big Sur coast. Nonetheless, the first impression is more likely, since Douglas-fir is by far the most common tree. The continual dominance of Douglas-fir is a noticeable by-product of the sunny, dry conditions; substantial old-growth stands and other factors indicate climax successional status in the open transitional and dry coniferous forest types of the San Juans.

We have divided San Juan forests into four main types, while recognizing that other variations and unusual plant associations occur. The first three forest types given correspond roughly to the order through which they would be passed moving from a southerly exposed slope to a northerly one. They bear close resemblance to those described by other authors.

OPEN TRANSITIONAL FOREST. Typically facing a southeasterly or southwesterly direction, this forest type seems to be intermediate between the south-facing meadow or rocky outcrop and the more densely timbered dry coniferous forest type. Correspondingly, the flora of the open transitional forest is a blend of open and forested habitat species but also is made up of forest-edge taxa and of some disturbed-site flora.

The open transitional forest, subject to higher winds and higher summer temperatures than other forest types, is most commonly found on moderately to very steep rocky slopes adjacent to salt water. Sites are well drained but soils are often thin and low in nutrients. Under these conditions, open stands of Garry oak *(Quercus garryana)* and Pacific madrona *(Arbutus menziesii)* can be expected, along with a sprinkling of Rocky Mountain juniper *(Juniperus scopulorum);* among trees, these species are the best indicators of the open transitional forest, although Pacific madrona also grows in the cooler, wetter forest, provided the site is well drained. Ocean spray *(Holodiscus discolor)* and Western serviceberry *(Amelanchier alnifolia)* are common shrub associates, as are the ubiquitous Nootka rose *(Rosa nutkana),* Common snowberry *(Symphoricarpos albus),* and shiny, handsome Tall Oregon-grape *(Berberis aquifolium).* The common understory tree is Scouler's willow *(Salix scouleriana),* although Hooker's willow *(Salix hookeriana),* described in the maritime section, is also readily found, but usually very near salt water. Douglas' maple *(Acer glabrum),* though not always common, will often be encountered here. Vines are represented by both Orange and Hairy honeysuckles *(Lonicera ciliosa* and *hispidula),* the former displaying uncommonly brilliant orange flowers in summer. Among grasses, Ryegrass *(Elymus glaucus)* and Yorkshire fog *(Holcus lanatus)* are overwhelmingly most numerous. Other indicator species are American vetch *(Vicia americana)* and Sierra Nevada pea *(Lathyrus nevadensis),* both displaying beautiful bluish or purplish flowers in season. Other typical species include Woodland tarweed *(Madia madioides)* and Pacific sanicle *(Sanicula crassicaulis).*

DRY CONIFEROUS FOREST. As the observer moves inland from the open transitional forest into the denser, pure forests of Douglas-fir, a slightly modified flora becomes evident. Gentler slopes, greater wind and sun protection, greater moisture, and heavier, more duff-laden soils are characteristic, although well-drained conditions are still present and forest conditions are still drier than in the moist mixed woodland. Salal *(Gaultheria shallon)* is the common shrub, but it seldom grows as densely or as large as it does on the outer coast. Low Oregon-grape *(Berberis nervosa)* and Little wild rose *(Rosa gymnocarpa)* are also common. The coppery haired Soopolallie *(Shepherdia canadensis),* a shrub more common east of the Cascades, is also widespread in dry island forests.

Interesting herb gardens may cover the forest floor, especially where a heavy moss cover is present. Fragrant bedstraw *(Galium triflorum)* and Broad-leaved starflower *(Trientalis latifolia),* with its six-sided white stars, plus Pathfinder *(Adenocaulon bicolor)* are very common. Yerba Buena *(Satureja douglasii),* a forest mint, and Twinflower *(Linnaea borealis),* a handsome matting species with strongly aromatic flowers, are frequent. Mountain sweet-cicely *(Osmorhiza chilensis),* Miner's lettuce *(Montia perfoliata),* and White-flowered hawkweed *(Hieracium albiflorum)* are also common. Where moss is heaviest, Rattlesnake-plantain *(Goodyera oblongifolia)* often turns up. Another orchid found here is Spotted coralroot *(Corallorhiza maculata).* As a saprophyte, this interesting purplish orchid needs no sunlight, deriving nutrients from decaying matter in the duff, and will thus grow in highly varied woodland sites, though it most commonly occurs where there is heavy shade.

MOIST MIXED WOODLAND. North-facing slopes, protected valleys, and poorly drained woodlands typically feature flora of the moist mixed variety. Cool, humid, windless conditions prevail, and forest floors are often rich in humus and are heavily shaded. The declining dominance of Douglas-fir indicates a new type of forest. Western hemlock *(Tsuga heterophylla),* Bigleaf maple *(Acer macrophyllum),* and other trees begin to be evident. At poorly drained sections, Red alder *(Alnus rubra)* and Western red-cedar *(Thuja plicata)* become more prevalent. Ferns are also more noticeable; among them Swordfern *(Polystichum munitum)* is the showiest and most common. Mossed-over logs and forest floors are the scene of much competition between various herbs, ferns, and tree seedlings, all crowding one another in an attempt to secure rare streaks of sunlight. Many herbs of the dry coniferous forest are present, but species like Siberian miner's lettuce *(Montia sibirica),* Western foam flower *(Tiarella trifoliata),* and Fringecup *(Tellima grandiflora)* seem especially confined to the moister, shaded areas. At poorly drained sites and valley bottoms, thickets of Salmonberry *(Rubus spectabilis),* Lady fern *(Athyrium filix-femina),* Large-leaved geum *(Geum macrophyllum),* Thimbleberry *(Rubus parviflorus),* and Coast red elderberry *(Sambucus racemosa)* are often evident. These species also form the understory of the Red alder forest.

DISTURBED WOODLAND. Any woodland with evidence of recent disruption would be included here. The toppling of trees by wind, fire, ocean-inflicted damage, or other natural havoc, aids the development of species associated with disturbed margins. However, the premier influence behind disturbed woodlands is man. Road cuts, housing developments, and the logging or thinning of forests all create disturbed situations. While many true forest species spread to disturbed edges, many European weeds also arrive, and often dominate. Not all of the flora typical of disturbed areas will be described here; there are enough species to legitimize a separate section, provided further on. In this section we describe some of the native species of the forest margin and of the thinned or cleared forest.

Several disturbed-site species are early successional types, like Red alder, which bring nitrogen to leached, disturbed soils. Many other species are weedy, such as Bracken fern *(Pteridium aquilinum),* Stinging nettle *(Urtica dioica),* various wonderful thistles *(Cirsium* sp.), and blackberries *(Rubus* sp.). In addition, geraniums (particularly *Geranium molle),* clovers (particularly *Trifolium repens),* and Oxeye daisies *(Chrysanthemum leucanthemum)* are common. Several native species creep from the woods to compete for space. Bitter cherry *(Prunus emarginata)* and Red-flowering currant *(Ribes sanguineum),* an early bloomer and favorite of hummingbirds, seem by far most common along forest margins. Other species include honeysuckle (both species), Nootka rose, Ocean spray, and Fireweed *(Epilobium angustifolium),* a widespread plant if ever there was one.

The four woodland types described include the greater part of the San Juan woodlands. Occasionally one finds exceptions, e.g., on Mt. Constitution, where old-growth, closed-canopy Douglas-fir–Western hemlock forests occur, which are virtually barren of understory flora—except for fall mushrooms, saprophytic heaths, and coralroots in spring and summer. This "bare" forest type is also present near Point Colville on Lopez Island. Another uncommon forest type, already mentioned briefly, is the pure or mixed alder woodland, characteristic of north-facing woodlands or of poorly drained situations, such as around lakes and marshes. The riparian vegetation associated with alder is described under the fresh water habitat section.

Lodgepole pine *(Pinus contorta)* plays a unique role in island woodlands. While dominant in few places, it grows in a myriad of far-flung habitats, with seemingly little preference pattern. Pure stands, about 20 feet or so tall, grow near the summit of Mt. Constitution; here, where fire raged some time ago, such a density could be expected in light of the tree's well-documented abilities to reseed after fire. Yet pure stands also overlook boggy sites on Hummel and Sportsmans Lakes on Lopez and San Juan Island respectively. Scattered individuals growing in open rocky slopes and sand dunes near the sea have earned it one of its other names, the Shore pine. It seems that Lodgepole pine is found over a wide ecological amplitude, occupying both ends of the moisture gradient, and when it colonizes, it often does so in pure stands.

FERNS AND FERN-ALLIES

LADY FERN *(Athyrium filix-femina)*. Among the first breaths of spring are the late March–April uncurling of Lady fern's bright green, fleshy fiddleheads. Soon huge, but weak, widespreading fronds up to 5 feet long will appear, seemingly overnight. This is one of the Pacific Coast's most abundant ferns, and it is common in the San Juans, although less so than at most lowland stations west of the Cascades, no doubt due to the islands' dryness. A typical species of damp places, Lady fern sometimes grows out of mud; shady woods, lake edges, ditches, etc., will also suffice. It may be confused with Mountain wood fern *(Dryopteris expansa)*, which has darker green symmetric fronds; they are also sturdier, feature greater leaflet incision, and linger through the winter. As the common name implies, it grows chiefly on decaying stumps or submerged wood, not in the disturbed moist-soils habitat of Lady fern.

SWORD FERN *(Polystichum munitum* var. *munitum)*. The long, trilling song of the mite-sized Winter wren in spring is often sung from a deep thicket of lush Sword ferns in one of western Washington's many wooded valleys. Nearly every gardener is acquainted with this species, as it is probably our best-known fern. Again, it seems somewhat less common in the islands than in most other places in western Washington, but it can be abundant (and normally is) in moist coniferous woods. Large and showy, the evergreen fronds reach 5–6-foot lengths under ideal conditions. The fronds are also densely chaffy, being covered with shiny brown scales. Also note the alternate, toothed pinnae. Now widely marketed, Sword fern has won acclaim for its attractive foliage and durability.

FLOWERS

White

SIBERIAN MINER'S LETTUCE *(Montia sibirica* var. *sibirica)*. In moist woods across Puget Sound country, one will often encounter a fleshy herb with immaculate white to pinkish flowers that seem to look back up at you as you examine them. In detail, the five petals of each flower will have thin reddish lines and will be shallowly notched. This delightful herb will prove to be none other than Siberian miner's lettuce, an excellent edible species if collected when young. Its appeal seems to be enhanced by its everlasting character: the blooms are apparent from April until as late as October, and during winter little basal rosettes of the long spoonlike leaves linger among the fallen leaves. Also called Western spring beauty and Candyflower, the latter name based on the neat striping, like candy cane, of the petals. To 12 inches tall.

MINER'S LETTUCE *(Montia perfoliata)*. One of the best wild salad greens in the San Juans, the foliage turning bitter only after considerable aging. Each of the succulent, clustered stems has one perfoliate leaf (i.e., a leaf surrounding the stem) just below the flowers, while the basal leaves come to an ovate or elliptic end. Small white flowers are present in spring. Two forms of this species are present in the San Juans: var. *perfoliata* is the common form of dry, open woods, growing in dense, luxuriant colonies. Plants are typically dark green and may reach 12 inches. Var. *glauca* is found on rocky bluffs and meadows, where it is usually solitary. Its foliage is sprawling and more tufted, and distinctly grayish yellow. This form is widespread but less common.

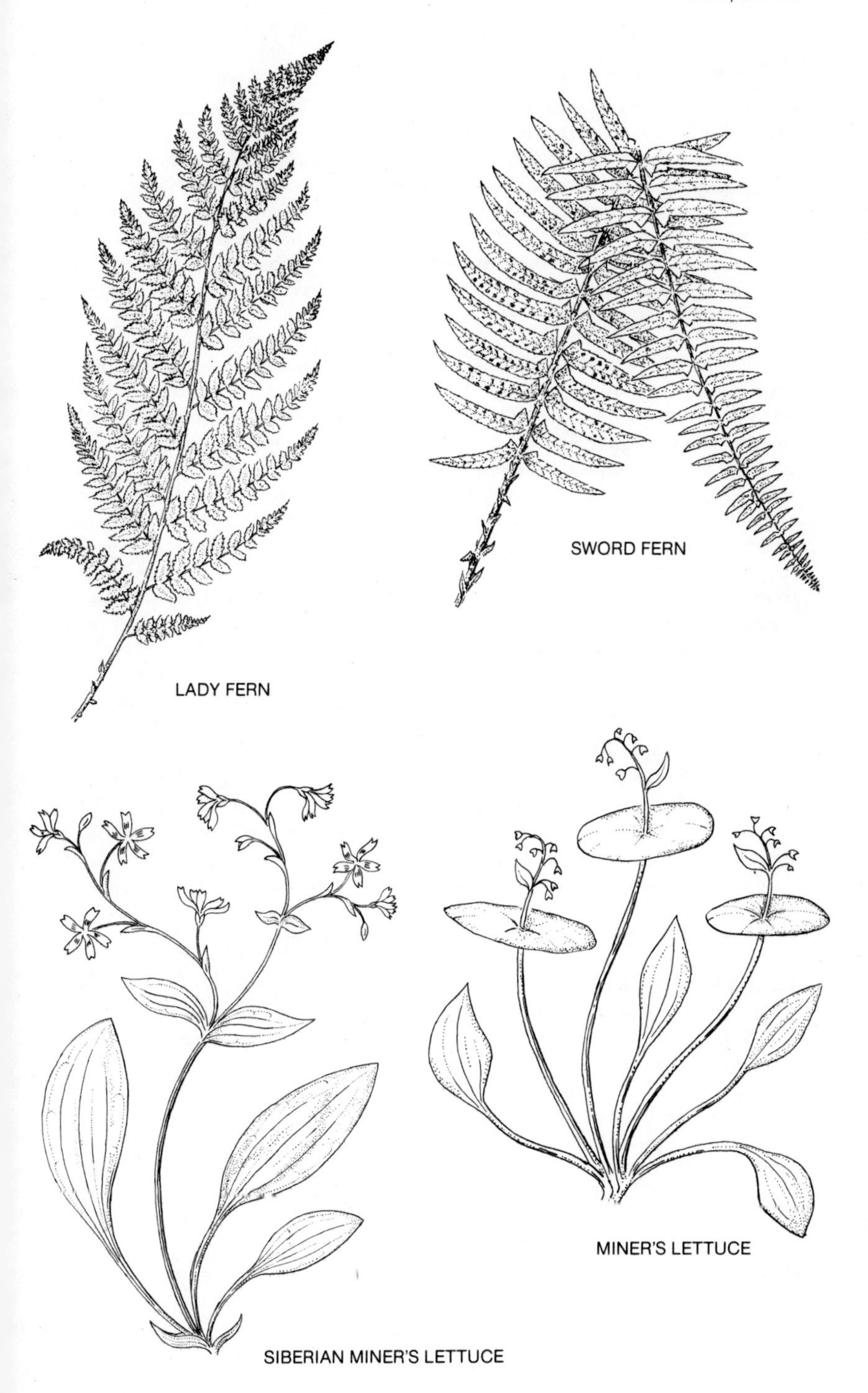
LADY FERN
SWORD FERN
SIBERIAN MINER'S LETTUCE
MINER'S LETTUCE

White (continued)

FRINGECUP *(Tellima grandiflora).* Don't mistake this plant for the similar Pig-a-back plant (Youth-on-age) sold occasionally in supermarkets, which, at least foliage-wise, Fringecup strongly resembles. Both are hairy and have maplelike leaves approximately 2–5 inches long. However, the cuplike flowers of Fringecup, with five white petals pinnately cleft in a fringelike, ragged manner, are quite different. The flowers bloom from the base upward and turn pinkish or reddish with age. Fringecup prefers damp, shady woods and valley bottoms, where it may reach 3 feet in height. Blooms from May to July, at which time it gives a sweet fragrance to the woodlands it graces. Common throughout the San Juans.

WESTERN FOAM FLOWER *(Tiarella trifoliata).* When walking through a dark, damp forest, watch for this delicate plant growing among other herbs on a downed mossy log. The fine spray of nodding, creamy white flowers is enchanting against the background of deep green forest-floor mosses and herbs during spring. Beauty is evident in each of the flowers, which feature long, white filaments of protruding stamens from each white cup. The overall appearance suggests that of alumroot, to which foam flower is closely related. Two varieties of Western foam flower, formerly considered separate species, occur in the San Juans. Although both have a single stem leaf and three leaflets per basal leaf, var. *trifoliata* has only shallowly lobed leaflets, while var. *laciniata* has deeply cleft, fully divided ones. Var. *trifoliata* is common throughout in damp, shady woods; var. *laciniata* is also widespread, but restricted to the dampest forests, especially on Orcas, Blakely, and San Juan Islands.

WOODLAND STRAWBERRY *(Fragaria vesca).* Although the berries are delicious, some people have suggested that the time expenditure in hunting and collecting the minute edibles exceeds the benefits derived therefrom. Whatever the case, most island residents should be well acquainted with this sturdy ground hugger, since it is very common in disturbed woodlands and margins. Pleasing to the eye are the white, five-petaled flowers of spring, which may be lost in the tangle of aerial stems and light green leaves. Two forms of Woodland strawberry occur in the islands: var. *bracteata* and var. *crinita,* with the latter most numerous. Interior strawberry *(F. virginiana)* is similar but has smooth, bluish green leaflets that lack the prominent venation and fleshiness of Woodland strawberry. Uncommon in more exposed habitats than those of *F. vesca.*

TRAILING BLACKBERRY *(Rubus ursinus).* Most people know of the "wild blackberry" or "dewberry," since its vines crisscross many forest floors and margins. Picking the excellent blackberries is an August treat for many, and everyone awaits the blackberry cobbler with wetted lips. This compensates for the annoying thorns, which will readily scratch or trip the unwary traveller. The five-petaled white flowers, similar to those of the other *Rubus* species, are obvious in spring and early summer when they sit nestled low among the dark or bluish green leaves. Becomes particularly aggressive in open, cut-over land, where the vines quickly overrun stumps and nearby vegetation.

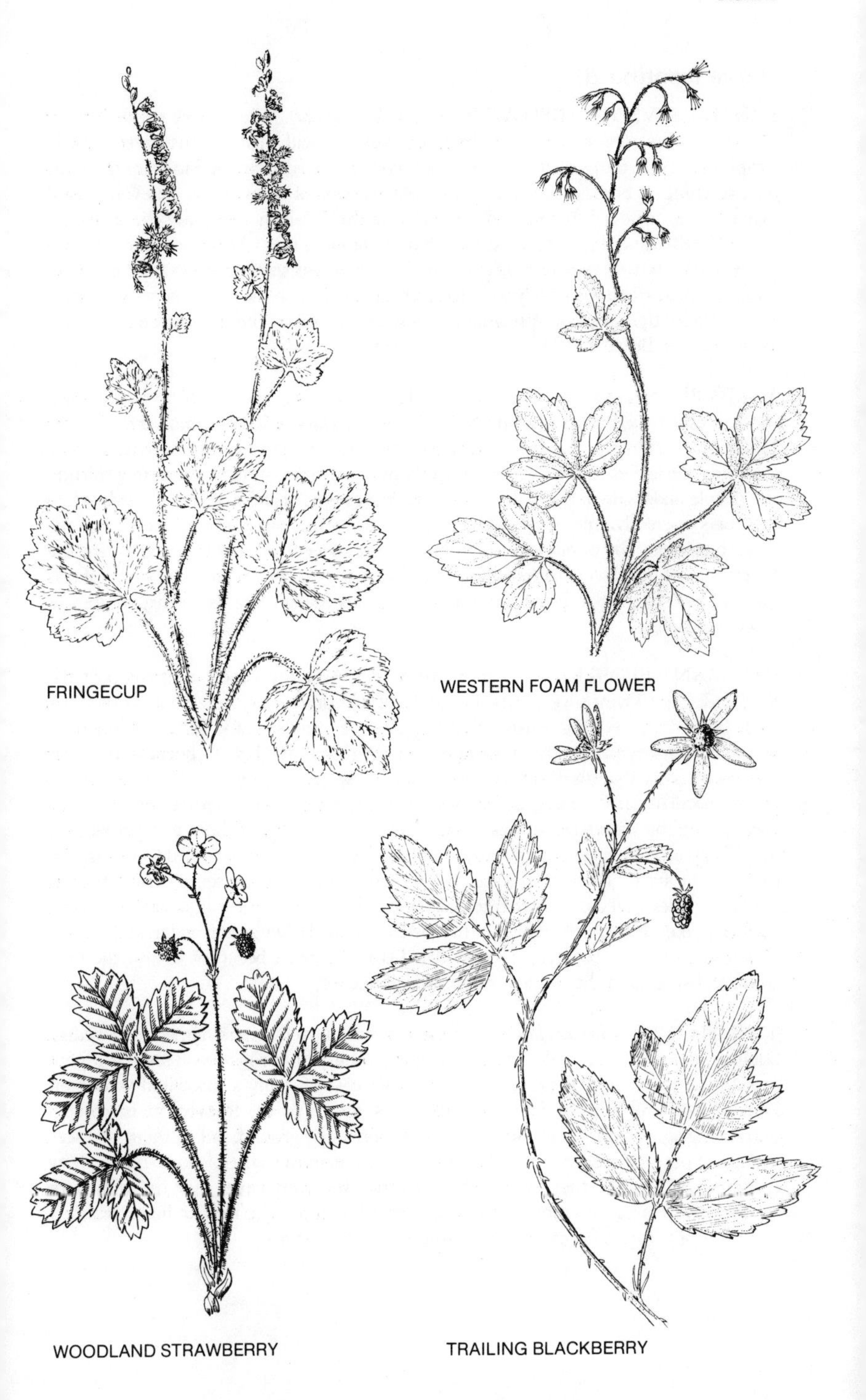
FRINGECUP
WESTERN FOAM FLOWER
WOODLAND STRAWBERRY
TRAILING BLACKBERRY

White (continued)

BROAD-LEAVED STARFLOWER *(Trientalis latifolia)*. With showy white "Star of David" flowers present in springtime, few persons will mistake this harmoniously simple species. Perhaps no forest wildflower is more abundant in the San Juans than this one, reaching in endless colonies across vast sections of forest floor. Unevenly sized ovate leaves, arranged in a terminal whorl atop the 3–5-inch stems, are characteristic. Some Northwest Indians ate the small but tasty tuberous roots, hence another common name, Indian potato. Northern starflower *(T. arctica)* has smaller leaves (to 2 inches in length) which are more loosely whorled and scattered along the stem; the white "stars" have rounded tips. Local on sphagnum bogs in the San Juans; the most extensive colony present at the Beaverton Marsh, San Juan Island.

YERBA BUENA *(Satureja douglasii)*. Tea lovers, here is a "good herb" for you. In fact, the Spanish name means good herb, no doubt due to its delicious aroma and pleasant flavor. No one having examined this little creeper up close can be immune to its charms. Crushing the leaves in one's hand releases a pure perfume; sometimes merely sniffing the purple underside of the little (½–1-inch) leaves is adequate to detect it. Little white trumpets discreetly emerge from the leaf axils in spring, often among the first of the forest bloomers. The woody stems, sometimes several feet long, are hairy and square, the latter trait common to all mints. Many will always remember the creeping stems of this familiar herb in the San Juans running across sunny forest margins and dry coniferous woodland floors.

FRAGRANT BEDSTRAW *(Galium triflorum)*. An elfin woods companion of the Pathfinder, the Twinflower, and other herbs of the coniferous forest floor is this small herb, often seen creeping waist-deep through a sea of mosses and lichens. Cinnamon scented when dried and sometimes when not, the tiny white flowers, borne in threes at the tips of axillary stems during mid-summer, are surprisingly potent for their size. As the vernacular name infers, bedstraws were employed to fill mattresses, although considering the small size, it must have taken a fair number of them to complete the job. This is an easy species to identify: as with all bedstraws, the whorled leaves, in this case five or six to a whorl, are the first tip-off. Next note the seeds, which feature hooked bristles well designed to catch on the passing fur of some animal or the exposed sock of a hiker. Finally, the leaves of this bedstraw are darker green, wider, and thicker than those of other bedstraws in the San Juans; Fragrant bedstraw is also the only woodland species in the islands, where it is abundant.

TWINFLOWER *(Linnaea borealis* var. *longiflora)*. Loved by and named for Swedish botanic genius Linnaeus, this delicate beauty of our coniferous woodlands is a gem among the forest flora, having won the admiration of many a woodland wanderer. During June and July the white or pink bells nod in paired fashion, giving off an entrancing scent from their perch 2–5 inches above the ground. Below them one sees little carpets of creeping, woody stems, often forming conspicuous mats. Why conspicuous? Because they bear numerous little evergreen leaves that are shiny dark green, nearly round, and remotely toothed. Found commonly in the San Juans, where it prefers mossy forest floors and sometimes the forest fringe.

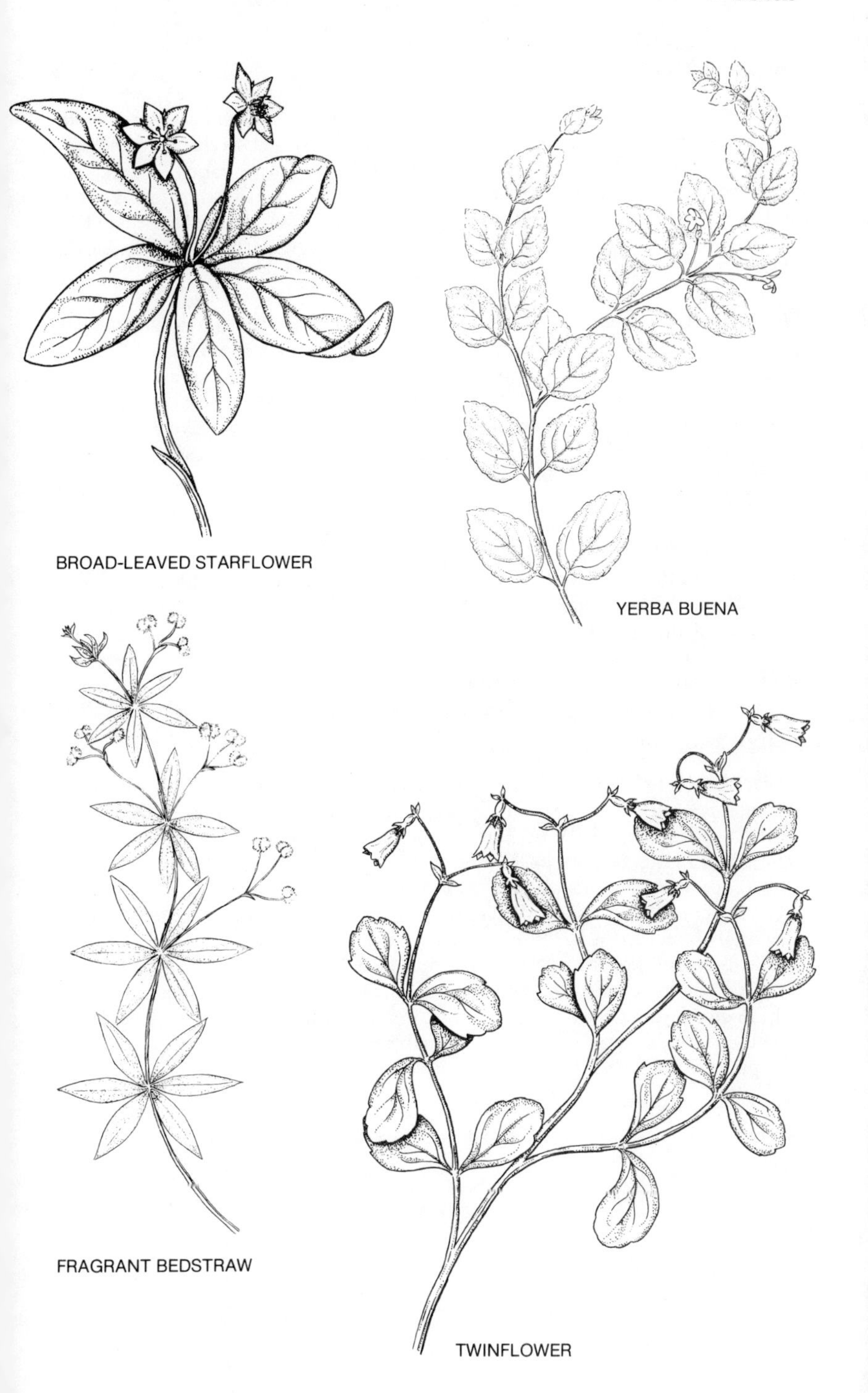
BROAD-LEAVED STARFLOWER
YERBA BUENA
FRAGRANT BEDSTRAW
TWINFLOWER

White (continued)

SCOULER'S HAREBELL *(Campanula scouleri)*. A lovely nymph in lavender-white, this wildflower could hold its own in any beauty contest. Your best chance of finding it is by visiting Orcas Island's mossy coniferous woods, particularly during June and July, although plants will be readily found on many islands. What is the secret to this fragile beauty? The outcurved, lily-white petals, which are sometimes lavender tipped, and the outshooting, pink-tipped pistil, which points downward. When not in bloom, note the heavily toothed leaves, which are mostly basal.

PATHFINDER *(Adenocaulon bicolor)*. An old friend of the deep woods is Pathfinder, or Trail plant, its large arrowhead leaves pointing this direction or that. The clear contrast between the upper leaf surface, which is dark green, and the lower, which has a distinct silvery sheen, gave rise to yet another name, Silver-green, and is denoted in the specific name, *bicolor*. During mid-summer, tiny whitish flowers are borne at the tips of 2–3-foot stems. These stems are hairy and, at the top, glandular (sticky), hence the genus name, *Adenocaulon* (aden, Greek for gland, and kaulos, stem). Abundant in the San Juans.

WHITE-FLOWERED HAWKWEED *(Hieracium albiflorum)*. Few species are more typical of the dry, open coniferous forest and margin than this one, so very common throughout the archipelago. Variably hairy, the leaves can be nearly smooth or so densely woolly that they appear grayish green. A late bloomer, usually showing its white flowers during July to September. 1–3 feet tall. This is the common *Hieracium* in western Washington and, interestingly, the only white-flowered species in Washington—the rest (there are a number), all found mainly east of the Cascades, are yellow flowered.

WHITE FAWN-LILY *(Erythronium oregonum)*. No announcement of spring is any more appreciated than that coming from this magnificent wildflower. One of the earliest of the early, the immaculate white or pinkish blooms are often in evidence by late March and frequently wither by May 1. No one will argue about the visual merits of the six outcurved white tepals (modified petals), regally centered by an open display of golden, pendant anthers, nor of the pair of glossy, thick, dark green leaves that are handsomely mottled in dark brown. Usually the flowers are borne singly on 12-inch or so stems, but occasionally twos and threes are found. The barrage of aliases is no doubt a reflection on the plants' popularity: Easter lily, Oregon fawn-lily, Oregon avalanche-lily, Dogtooth violet, and others are available for those who prefer another name. Fairly common in dry open woods and meadows throughout the San Juans, although it has disappeared from many places, due to picking and habitat destruction.

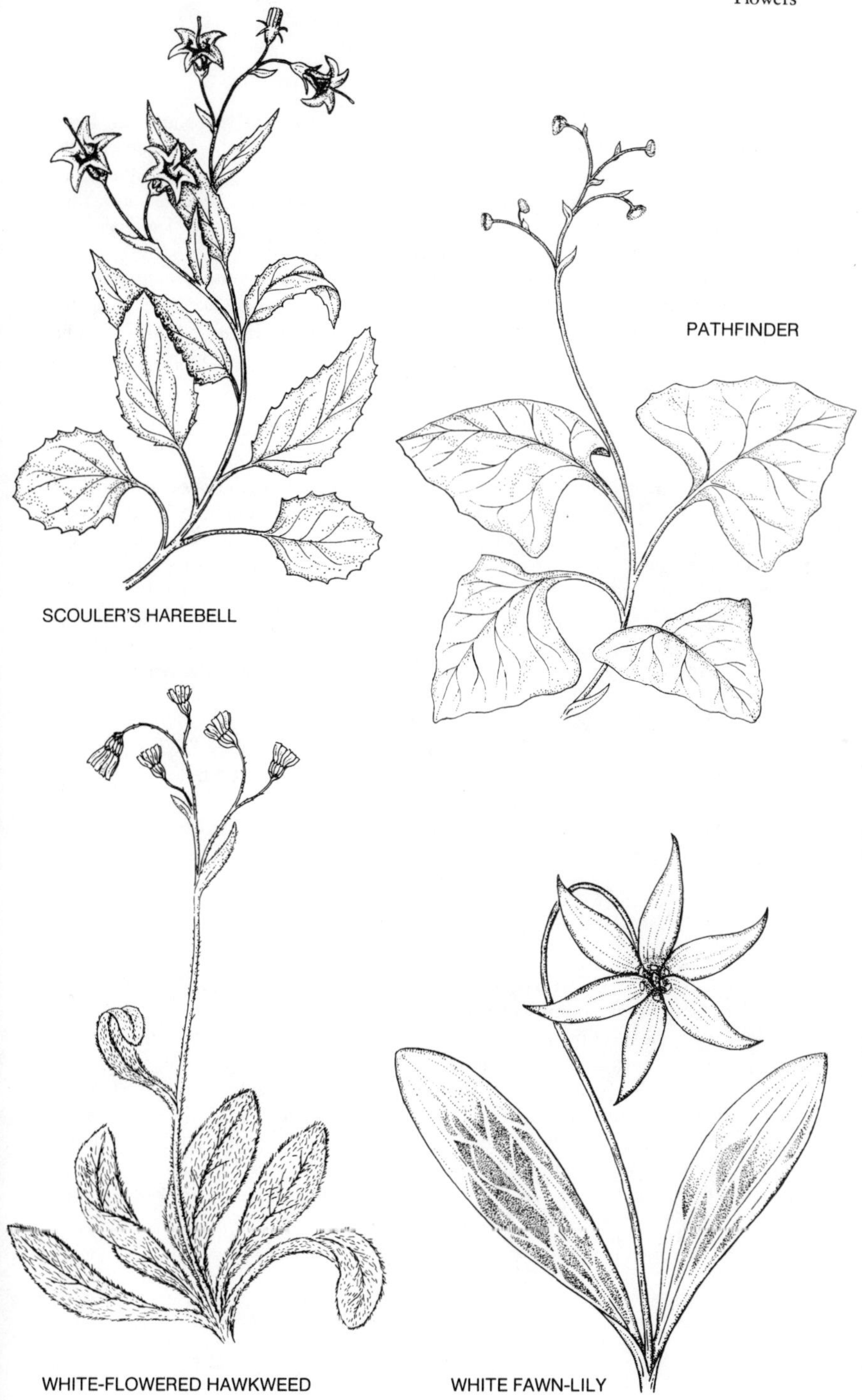
PATHFINDER
SCOULER'S HAREBELL
WHITE-FLOWERED HAWKWEED
WHITE FAWN-LILY

White (continued)

RATTLESNAKE-PLANTAIN *(Goodyera oblongifolia).* The coniferous forest floor—little herb gardens in light green, fallen twigs blanketed by a spider web, the web in turn sprinkled by yellow-green fir needles, mossed-over logs with seedlings, and the dark green, white-streaked leaves of Rattlesnake-plantain squeezed in a tight rosette between a carpet of mosses—that's the coastal woods scene. The evergreen leaves of the Rattlesnake-plantain, whether heavily mottled in milky white or with a simple white midvein, are well camouflaged, so they may be overlooked, although this species is common. Among a family of champions, the August blooms of this orchid are a disappointing, dull greenish white, rising 12–18 inches off the ground. The genus name is not a misspelled recognition of Goodyear, the tire tycoon, but correct for William Goodyer, an English botanist (1592-1664).

Yellow

PACIFIC SANICLE *(Sanicula crassicaulis* var. *crassicaulis).* An indicator species of the dry, transitional forest found on rocky slopes above salt water. The palmately lobed, toothed basal leaves, with their light venation, are diagnostic. May and June are the months to see the umbellate dark yellow flowers, borne on tough 1-2-foot stems. This herb densely populates certain parts of English Camp, San Juan Island. Purple sanicle *(S. bipinnatifida)* has unevenly arranged, showy purple flower heads, blooming at roughly the same time, and has more deeply cut leaves (one–two times pinnatifid). Uncommon; mostly grows in dry, gravelly meadows on San Juan and Lopez Islands.

WOODLAND TARWEED *(Madia madioides).* Displaying rich golden yellow flowers during the summer months, this species has little difficulty establishing itself as most attractive among six *Madia* species occurring in the San Juans. Each flower has eight trilobed petals that sit flat atop a sticky-haired receptacle bract. Before and after blooming, watch for the long (to 10 inches), narrow, hairy leaves, which clasp the stem. Madi, the Chilean name for Chilean tarweed *(M. sativa)*, is the origin of the genus name. Common in dry, open, Douglas-fir woodlands, usually within sight of salt water.

Orange

ORANGE HONEYSUCKLE *(Lonicera ciliosa).* Dazzling orange to rarely scarlet, the spring–early summer blooms will warrant a second glance. A denizen of the dry, open coniferous forest, this honeysuckle often enjoys climbing 15–20 feet up the trunk of Douglas-fir, where the showy flower clusters look downward from their perch on a young bough. A distinctive feature is the single connate (fused) terminal leaf, below which are many paired, grass-green, short-hairy (i.e., ciliate, hence the specific name) leaves. The cordlike, tough vines routinely reach 20 feet or more. Pollinated almost exclusively by Rufous hummingbirds, Orange honeysuckle has no need for a fragrance to attract insects and thus has evolved without one, unlike some of its close relatives.

RATTLESNAKE-PLANTAIN
PACIFIC SANICLE
WOODLAND TARWEED
ORANGE HONEYSUCKLE

Orange (continued)

COLUMBIA LILY *(Lilium columbianum)*. A "wow!" may be uttered as someone locates this behemoth of a lily, surprising in its size (3–5-foot tall stems) and in the splendor of its blooms. Fiery orange and several inches wide, its nodding flowers can often be spotted along a roadside, even from a speeding car. Closer observation reveals each of the three–nine flowers to have six pinned-back tepals that are splashed with maroon freckles, below which hang the brownish purple anthers. A forewarning of this June–July spectacle is the upcoming stems, which feature even whorls of six–nine smooth leaves. Usually solitary or in small groups, Columbia lily is not very common but is widespread in the islands. Watch for it along gravelly roadsides skirting dry to moist woodlands, where it often becomes dust covered along with other marginal species; also found in open coniferous forests, especially those that have been thinned or otherwise disturbed, preferring exposed margins.

Pink

PIPSISSEWA *(Chimaphila umbellata* var. *occidentalis)*. Growing in the dark, damp, moss-laden world of the deep coniferous forests, this little wildflower is widespread in, and west of, the Cascades, especially at middle elevations. The melodic Indian name seems somehow appropriate and should be retained ahead of the other name, Prince's pine. Any hiker is bound to break stride at the sight of the nodding flowers, which are waxy, soft white or pinkish, and enchanting. Winter-loving, the meaning of the genus name, refers to the evergreen foliage, which is arranged in uneven whorls of toothed, leathery leaves, a good field mark when the plant is not blooming (all year except late June through mid-August). Little prince's pine *(C. menziesii)* is smaller (to 4 inches tall as opposed to the 6–12-inch height of *C. umbellata)*, fewer flowered (one–three versus five–fifteen with *C. umbellata)*, and has smaller, dark bluish green leaves, often with weak, whitish midveins. Watch for both species on Mt. Constitution, where they are most common.

PINK WINTERGREEN *(Pyrola asarifolia)*. A glowing pink inflorescence, rising above a carpet of light green herbs and deep green mosses, brings a startling elegance to the dim hue of the forest floor during June and July. An agreeable balance is apparent in the well-ordered, alternate arrangement of waxy, nodding blooms. Each flower is about ½ inch wide and has a well-defined, hooked style. When this plant is not flowering, an observer intent on other plants can easily overlook the lustrous leaves, which suggest the young foliage of the Salal growing nearby. Largest of the *Pyrola* group, it may reach 16 inches in height. Like many deep-woods species, this herb is partially saprophytic, living off decaying plant material in the duff. The need for chlorophyll-receptive leaves is thus negated and, in fact, nearly all of Washington's wintergreens have leafless forms (var. *aphylla)*. In the San Juans, the leafless form is more widespread, though it is scattered and often solitary. The leafy form (var. *asarifolia)* is locally common on Mt. Constitution but rather scarce elsewhere in our area.

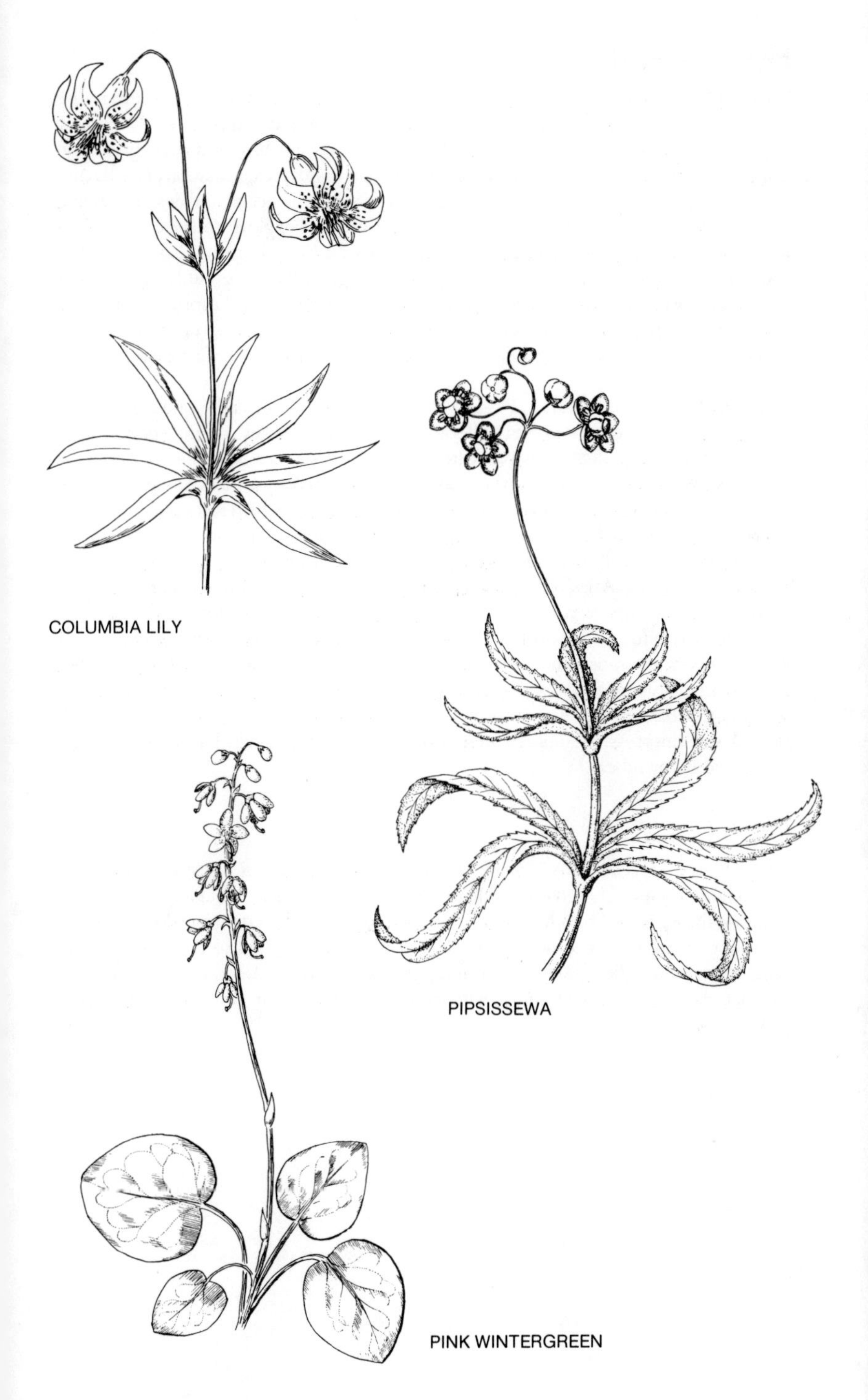
COLUMBIA LILY
PIPSISSEWA
PINK WINTERGREEN

Pink (continued)

FOXGLOVE *(Digitalis purpurea).* Only the most heedless wayfarer could fail to notice the prolific blooms, their whites, roses, pinks, and purples now flourishing along many wooded roadsides and margins west of the Cascades. Astonishment would be a logical response to the evolutionary wonder the huge flowers display; examination of each hairy tube reveals white spots on the lower throat, these being the nectary guides or "landing gear" to aid each bumblebee in making a safe landing. Foxglove contains digitalis, a strong stimulant used by those suffering from heart ailments; it can also be lethal to those who don't need it. Often very large, plants may grow 8 feet high, although 4–5 feet is average; the hairy, coarse leaves are often over 12 inches long. Common in the San Juans, this is a premier species of the disturbed woodland, showing up in thinned or bulldozed woodlands, cut-over hillsides, forest edges, roadsides, etc. European.

CALYPSO *(Calypso bulbosa).* Passing through a heavily shaded, coniferous forest during May, the watchful traveller will sometimes be rewarded by coming upon a classic orchid of bewitching elegance, so delicate as it grows alone on a bed of heavy humus or fern moss. Nature's artistry is hard to miss in the perfect design—a nodding head with five rosy-purplish tepals (modified petals) above a little face, which has a pink upper lip, a white lower one flushed with pink, and a blood red streaked throat that seems as if it were hand painted. If that isn't enough, the heads often have a pleasing, unusual fragrance. Calypso, Atlas' daughter in Greek mythology, was a sea nymph, often concealed below the waves, and here her name honors the lovely charms and inconspicuous nature of our orchid. The roots grow in association with certain fungi and, thus, attempts to transplant are futile; plants also cannot tolerate picking or other disturbances. This species was formerly believed to be rare in our area, but naturalists in Washington have now found that it is widespread. The San Juans are particularly blessed, with impressive colonies present on many islands, such as Jones, Sucia, Lopez, Orcas, and many others.

Purple

AMERICAN VETCH *(Vicia americana).* Here is the coniferous forest representative amid a number of vetches in the San Juans. The majority of them, including this one, are common throughout. From May to July, the dark reddish to bluish purple flowers bloom in groups of six–ten. At other times watch for four–eight pairs of pea green, spine-tipped leaflets and the twisted, often zigzagging stems. Vars. *truncata* and *villosa* occur with us. An excellent wild edible with both agreeable flavor and substantial nutritional value.

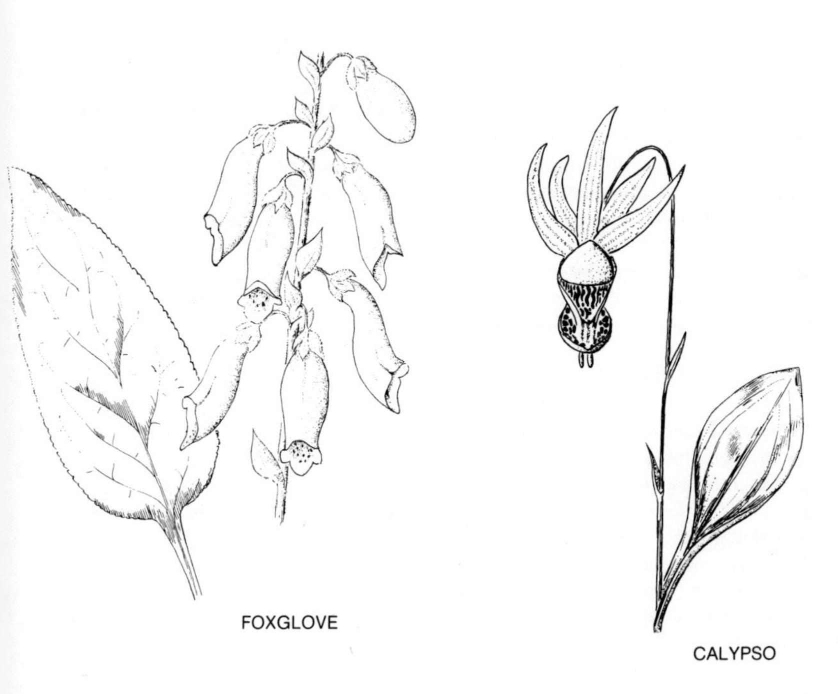

FOXGLOVE

CALYPSO

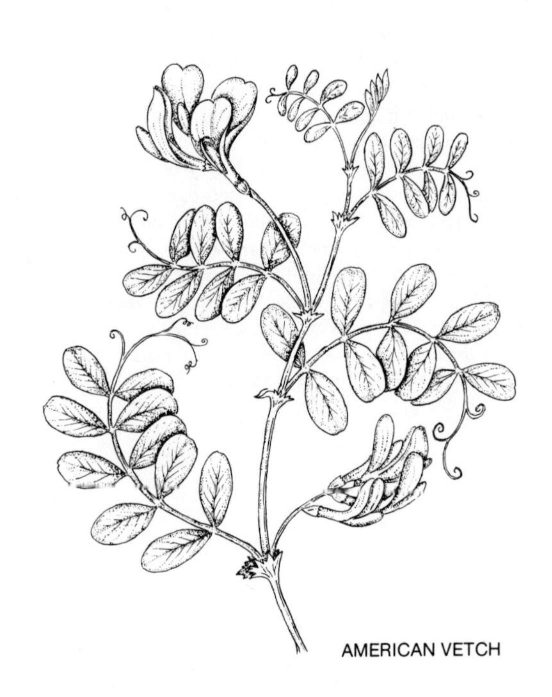

AMERICAN VETCH

Purple (continued)

HAIRY HONEYSUCKLE *(Lonicera hispidula).* At the Lopez Island ferry landing, in the open woods above salt water, one is assured of finding this honeysuckle, as one can in any sloping Douglas-fir forest above salt water throughout the archipelago. Smaller and less celebrated than Orange honeysuckle, this vine has smaller, leathery leaves that vary in color from a dull purplish green to grayish blue. Blooms during June and July; the flowers, although less eye-catching than those of Orange honeysuckle, are sometimes striking, varying from dense purple to a soft pink, and are usually yellow centered. A few bright red berries usually appear following the bloom in late summer. Like those of Orange honeysuckle, they are very bitter.

SPOTTED CORALROOT *(Corallorhiza maculata* ssp. *maculata).* An intriguing saprophytic or parasitic denizen of deep woods, this coralroot can thrive on dead, dying, or living tissues of other plants, and when it parasitizes, conifers are invariably the victims. Whatever its food source, green, chlorophyll-collecting foliage is unnecessary and, accordingly, plants have no real leaves. Instead, a reddish purplish spike bears scattered, fleshy scales, all that remains of the long-lost leaves. Blooming from May to July, the flowers are striking: each has a white lower tip with purple spots. By July, the stalks fade to brownish and support hanging, round seed cases. Common in the islands, often growing under the darkest, densest of forest canopies. A rare yellowish form, presumably the same one that occurs rarely on southern Vancouver Island, is reported from Orcas Island by Dr. Frank Richardson.

Green

MOUNTAIN SWEET-CICELY *(Osmorhiza chilensis).* Fernlike best describes the leaves, and a nuisance is a most apt definition for the black, spiny seeds, which have an annoying habit of lodging themselves in one's socks and then poking the skin with every footstep. However, the sweet-tasting root is a reputed aphrodisiac, so the plant can't be all bad. Between April and June, finding the tiny clusters of greenish flowers can be difficult, but from June to August locating the seeds is a cinch. If they haven't already found your socks, they will be obvious at the tip of a 1–3-foot stem, which lingers prominently after bearing the seeds. Both the common name (cicely) and the genus name *(Osmorhiza)* are of Greek origin, the latter meaning scented root. One might say that this sweet-cicely grows quite nicely in precisely those woodlands with ample shade, especially coniferous ones. Abundant in the San Juans. Purple sweet-cicely *(O. purpurea)* is very similar but has a broad stylopodium (the pointed appendage at the top of the seed), not spikelike, as with *O. chilensis.* The seeds are spiny only halfway up and the flowers vary from purplish to greenish. Local in the San Juans, occurring in dense coniferous forests, chiefly on Orcas and Blakely Islands.

HAIRY HONEYSUCKLE

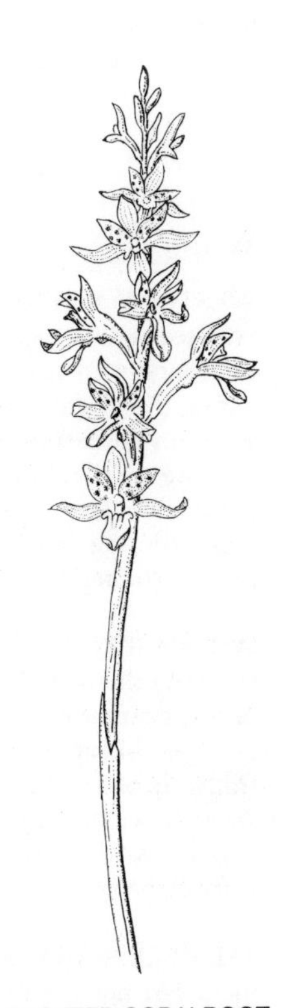

SPOTTED CORALROOT

MOUNTAIN SWEET-CICELY

Green (continued)

ELEGANT REIN-ORCHID *(Habenaria elegans).* The refined flower detail of many orchids, like this one, is most pleasing. Few San Juan orchids are as ubiquitous as Elegant rein-orchid, which has been successful enough to have colonized some of the roadside ditches on San Juan Island. Generally a dry, open coniferous forest species, it has the notoriety of being one of a few herbs that can grow under a madrona canopy. Elegant rein-orchid may be confused with the San Juans' other four rein-orchids but not if the observer remembers some simple characteristics. First, the flowers are greenish and bloom in open clusters. Second, the spur (the long appendage below the petals) is white and ¼–½ inch long. The leaves, if present, are basal, long, and usually lanceolate or linear.

SHRUBS

White

WESTERN SERVICEBERRY *(Amelanchier alnifolia* var. *semiintegrifolia).* Often browsed by deer, this shrub is frequently low (2 feet or so tall) and matted, though under ideal conditions it may grow to 15 feet. Bright white flowers decorate the branches during April and May, drawing attention at their forest borderland perch. The summertime berries are red or blackish and of low quality, at least in our area; however, they are devoured by birds and other creatures and were an element of the Indians' pemmican. They were also a staple food source of Canadian citizens in the prairie regions during the deprivations of the 1930s, hence another name, the Saskatoon berry. A fairly common shrub in the San Juans along open, rocky woodlands.

OCEAN SPRAY *(Holodiscus discolor).* Abundant from one hillside to the next forest, this is another shrub that can be found everywhere in the islands. Its ubiquitousness has perhaps damaged its image, so we'll review its amiable features. Most pleasant are the June–July blooms, their creamy color and downward-curling manner suggestive of the frothy tip of a crashing wave, hence the name. The soft little leaves (1–3 inches long) are mellow, and they make a pretty sight as they wave in the summer wind. Finally, the durable wood of the sprawling stems was used by olden-days woodsmen for finish work carpentry.

THIMBLEBERRY *(Rubus parviflorus).* Damp, shady woods and sometimes dry, open forests are home to this well-known Northwest shrub. Suggestive of maple, the hairy, thick leaves are palmately lobed and may be 8 inches wide. During May, conspicuous white flowers rise above the foliage, lasting until July when the fruits develop. Although they may be seedy and bitter, the burgundy to purplish berries are frequently very sweet and tasty, hence their popularity with many forest creatures and some humans. Thimbleberry invariably forms junglelike, 2–4-foot thickets with other species at poorly drained, shady sites throughout our area.

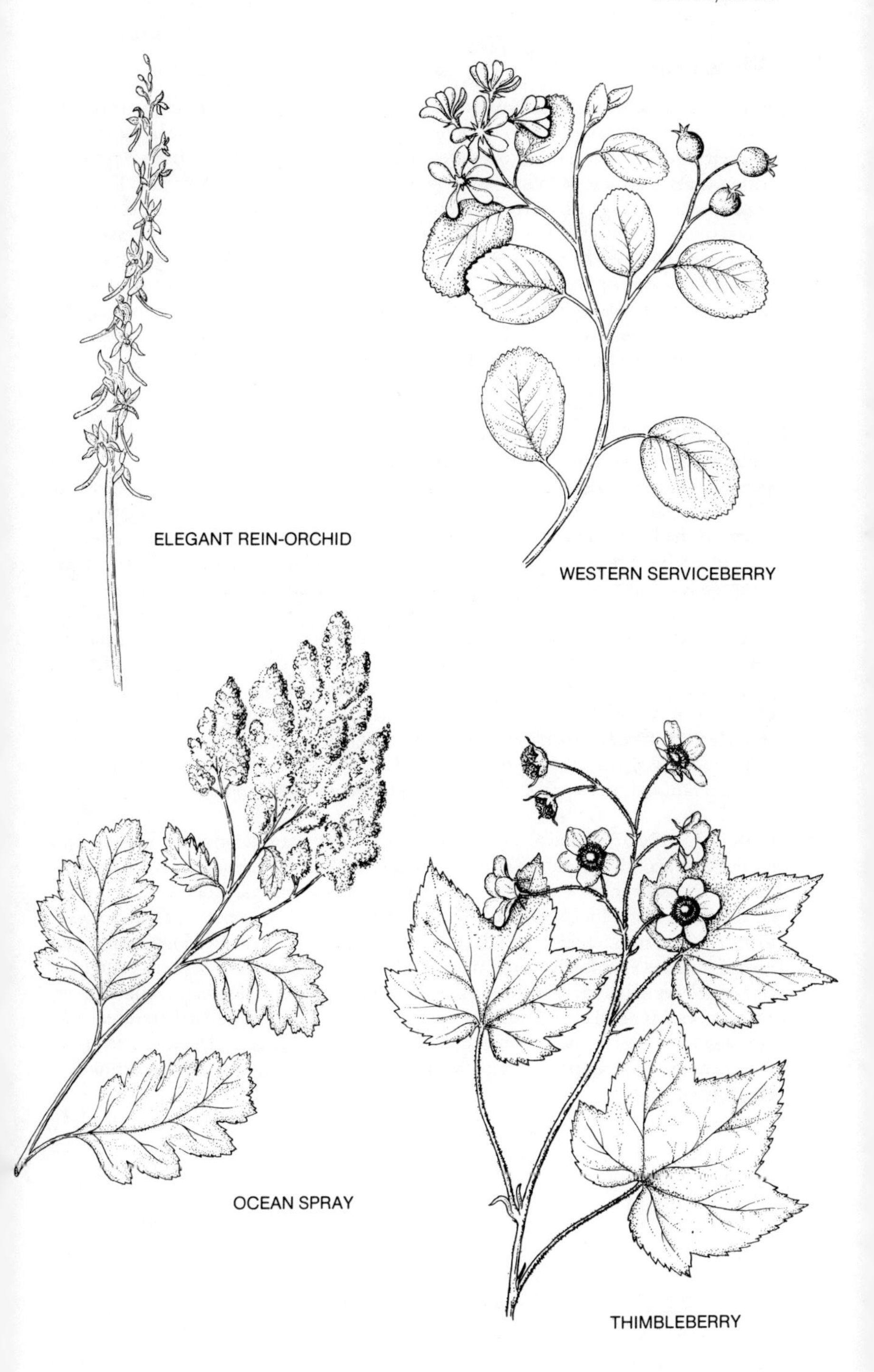
ELEGANT REIN-ORCHID
WESTERN SERVICEBERRY
OCEAN SPRAY
THIMBLEBERRY

White (continued)

SALAL *(Gaultheria shallon)*. An easy habit to fall into would be to take this attractive shrub for granted, since it is so ubiquitous throughout our coastal forests. It may be necessary at times to remind ourselves of its redeeming qualities. Beginning with spring characteristics, we witness the pendant, white to pinkish bells that spread a sweet scent through the forest understory, attracting bumblebees that constantly attend them. At the same time the new, light green leaves form striking contrasts with the older, dark ones below. By July, black berries in dense groups weigh down the branches; their taste may be sweet and, although they are mealy, some people like them. In winter, Salal teams up with evergreens to ward off the gray monotones of the now-barren deciduous plants. The message has not been missed by landscapers and florists, who commonly use the shrub. Northwest Indians gave Salal its name; the dried fruit was also a staple of their diet.

COAST RED ELDERBERRY *(Sambucus racemosa* var. *arborescens)*. Wine-red and gleaming in the late summer or autumn sunlight, the dense berry clusters of Coast red elderberry are well-known to most residents west of the Cascades. Flourishing in damp, low ground, often where shady, this stout shrub (to 20 feet) is often among the first plants of the humid valleys to give a sneak preview of spring's light greens, joining Lady fern's fiddleheads in unfurling its large leaflets. Not long afterward, the huge, creamy white flower clusters spread a rather rank, pungent odor through the area. The berries are a favorite of Band-tailed pigeons, Robins, and Swainson's thrushes. The spreading canes are bumpy and pith filled on the surface. Older stems are substantial enough to collect mosses and lichens and, indeed, several interesting mushroom species are specific to them, including certain members of the deadly *Galerina* group. Occasionally one finds plants on open, rocky slopes, and in this habitat one may also stumble upon Blue elderberry *(S. cerulea)*, although in the San Juans it is rare, occurring scatteringly along the west side of San Juan Island. It has pale blue berries, present often into late fall, and flat-topped, white flower clusters during summer.

COMMON SNOWBERRY *(Symphoricarpos albus* var. *laevigatus)*. This species battles with Nootka rose and Salal for the honor of the most abundant San Juan shrub, its weak 2–3-foot thickets usually present on dry, exposed slopes, in dense shady woods, at roadsides—or almost anywhere! Unfortunately, it is rather difficult to gain inspiration writing about the foliage and blooms: the former are nondescript, dull greenish leaves (1–2 inches long) that are often irregularly lobed, while the latter are tiny pink bells that arise from the leaf axils during June–July. The creamy white berries are bitter and generally avoided by birds. However, they are a familiar, agreeable sight of many roadsides in late summer and linger well into fall or early winter, when on damp, chilly days, they provide a memory of summer's bounty against a background of gray, brittle twigs.

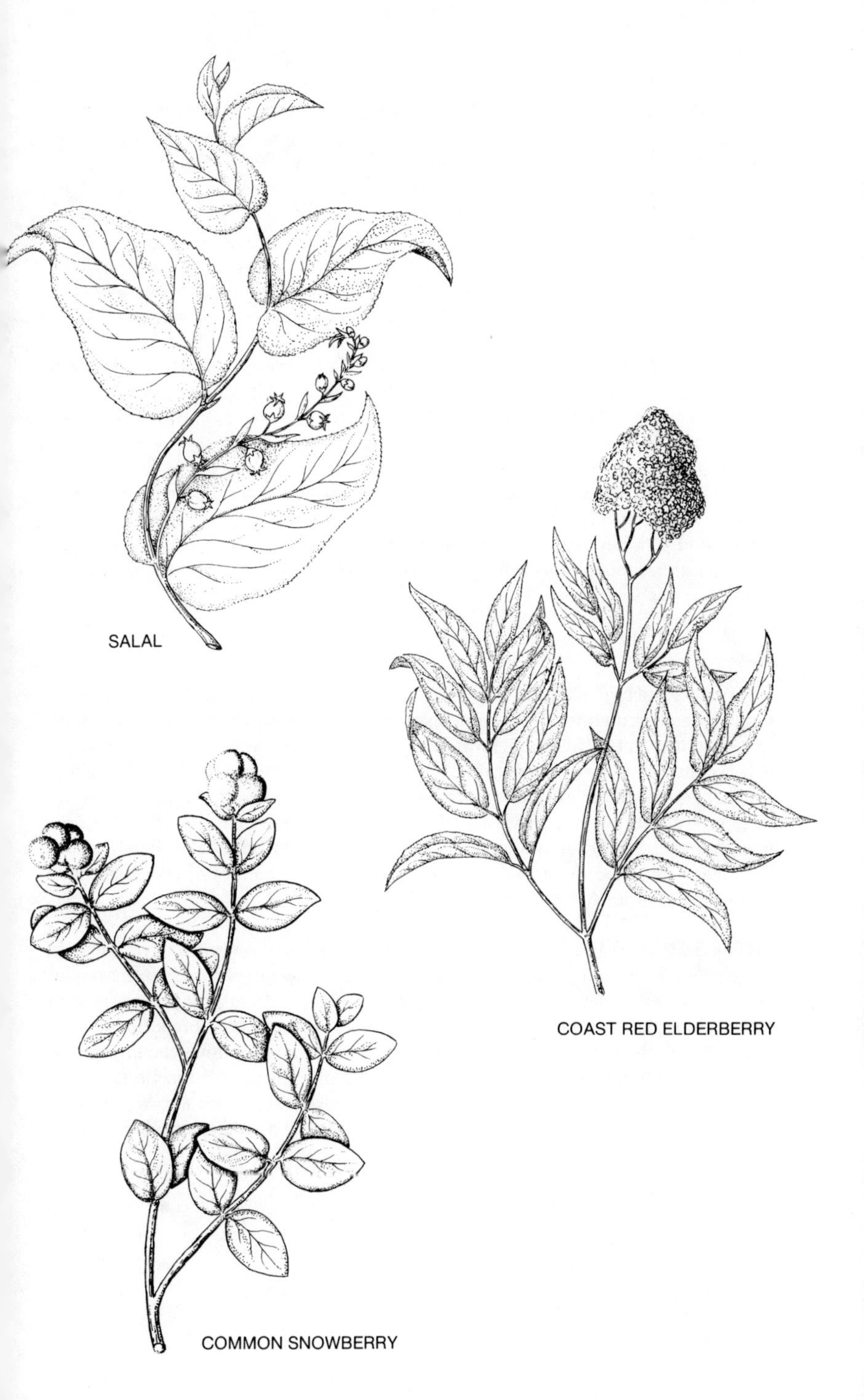
SALAL
COAST RED ELDERBERRY
COMMON SNOWBERRY

Yellow

TALL OREGON-GRAPE *(Berberis aquifolium)*. More visually appealing and taller (to 5 feet or more) than Low Oregon-grape, this evergreen shrub displays thinner leaves, whose shininess makes them seem polished or wet. The latter likeness inspired the specific name, *aquifolium,* or water leaf. Prefers more exposed sites, like open rocky slopes near salt water or forest margins. Northwest Indians derived an excellent yellow dye from the striking yellow inner bark and lemon-yellow spring flowers. The berries, although rather unsavory eaten raw, have been used in jams and jellies. A desirable ornamental because of its heartiness and attractive leafage, which sometimes becomes a rich scarlet along the edges just before being discarded as the new foliage replaces it.

LOW OREGON-GRAPE *(Berberis nervosa)*. There would be little difficulty in imagining this sturdy plant to be a relative of holly, as the 9–21 leathery, prickly leaflets suggest elongate, flattened holly leaves. However, there is no relationship and, in fact, in many respects they are quite different. The full green foliage lacks the shine of holly, that absence having inspired a second, less than complimentary name, Dull Oregon-grape. Another foliage characteristic is the leaf venation, acknowledged in the specific name, *nervosa.* In spring, light yellow, faintly fragrant flowers rise above the leafage, followed in summer and fall by the pearly blue berries, which are edible. Abundant, forming extensive colonies in shady woods by use of rhizomes. The woody stems reach 2 feet.

Red

RED-FLOWERING CURRANT *(Ribes sanguineum)*. By late March, the scarlet or pinkish pendant blooms of this shrub exuberantly announce spring's arrival, and the first incoming Rufous hummingbirds, hungry for the sweet nectar within, always seem to arrive just as the first blooms appear. In the San Juans, open woods and dry, sunny forest margins are this plants' preferred habitat. After the blooms wither in May, the dull, grayish green leaves may be easily missed when showier foliage is present. An open-branching bush that sometimes reaches 5 feet in height; also called Blood currant. The blue to black berries are of variable quality but are generally of less edible value than those of other currants.

MOUNTAIN LOVER *(Pachistima myrsinites)*. Sucia Island's wooded rocky slopes are an excellent location in which to find this handsome evergreen shrub en masse, but there are several other places one can expect it. Thick, dark, glossy green leaves are paired along the branches; their attractiveness has won the attention of florists, who collect boughs for flower displays. From mid-spring to early summer the tiny reddish brown flowers emerge from the leaf axils, giving off a pleasant odor. Prefers semi-open, often steep sites; most common in the Cascades and Olympics in Washington. The frequently used name, False box, is based on the plant's similiarity to Box *(Buxus)*, cultivated for garden hedgerows.

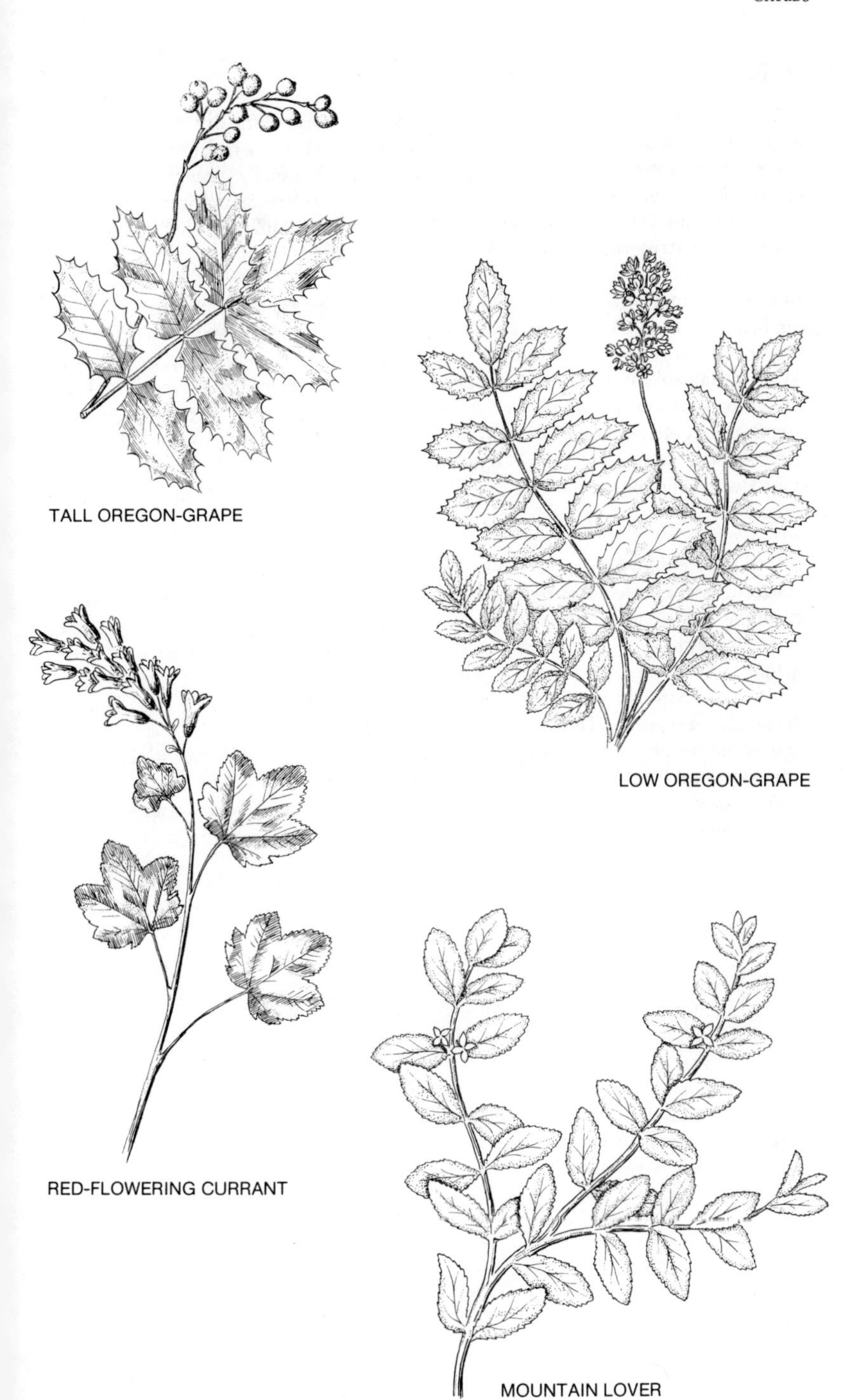
TALL OREGON-GRAPE
LOW OREGON-GRAPE
RED-FLOWERING CURRANT
MOUNTAIN LOVER

Pink

LITTLE WILD ROSE *(Rosa gymnocarpa)*. An old friend of the dry or moist shaded woodland, this rose brings cheer in spring when its golden-stamened, lush pink flowers bloom. This is the smaller of our two common roses, usually growing 2–4 feet in height. The leaves are small, light green, and doubly serrate, and the stems are evenly armed with small but sharp bristles. Often grows solitary, unlike the Nootka rose. The fruits are small, varying in color from orange to reddish.

NOOTKA ROSE *(Rosa nutkana* var. *nutkana)*. Perhaps the commonest shrub in the San Juans, forming dense 4–6-foot thickets along salt water, roadsides, forest edges, and exposed places generally. The showy pink flowers may be 3 inches or more wide and usually grow alone at the branch tips. The rose hips, appearing in late summer and fall, are an excellent source of vitamin C and can be used to make a tasty tea or syrup; or they may be eaten raw, although they are seedy and tart. Originally found and described from Nootka Sound, Vancouver Island, hence the name. After a spring shower, the thin, dew-laden pink blooms sometimes fill the air with an exquisite fragrance not soon forgotten.

SALMONBERRY *(Rubus spectabilis)*. During a mild February, the observer may be amazed to find a perfect rosy red or pinkish starflower of this thorny plant, long before any of the first leaves have unrolled. By March and April, more blooms will be present, although they may become inconspicuous as heavy leafage unfurls, hiding them below. By June and July the tangerine-orange to rosy purplish berries are being gobbled up by Robins and many other animals; most people enjoy the soft, rather large berries, because they are juicy and often sweet. In addition, Indians ate the peeled young shoots; their common practice of eating salmon with them is the apparent origin of the common name. Dense thickets of Salmonberry to 8 feet in height are not uncommon in shaded valleys and poorly drained places in the San Juans. A notable characteristic is the typical peeling and flaking of the canes, which also turn golden with age.

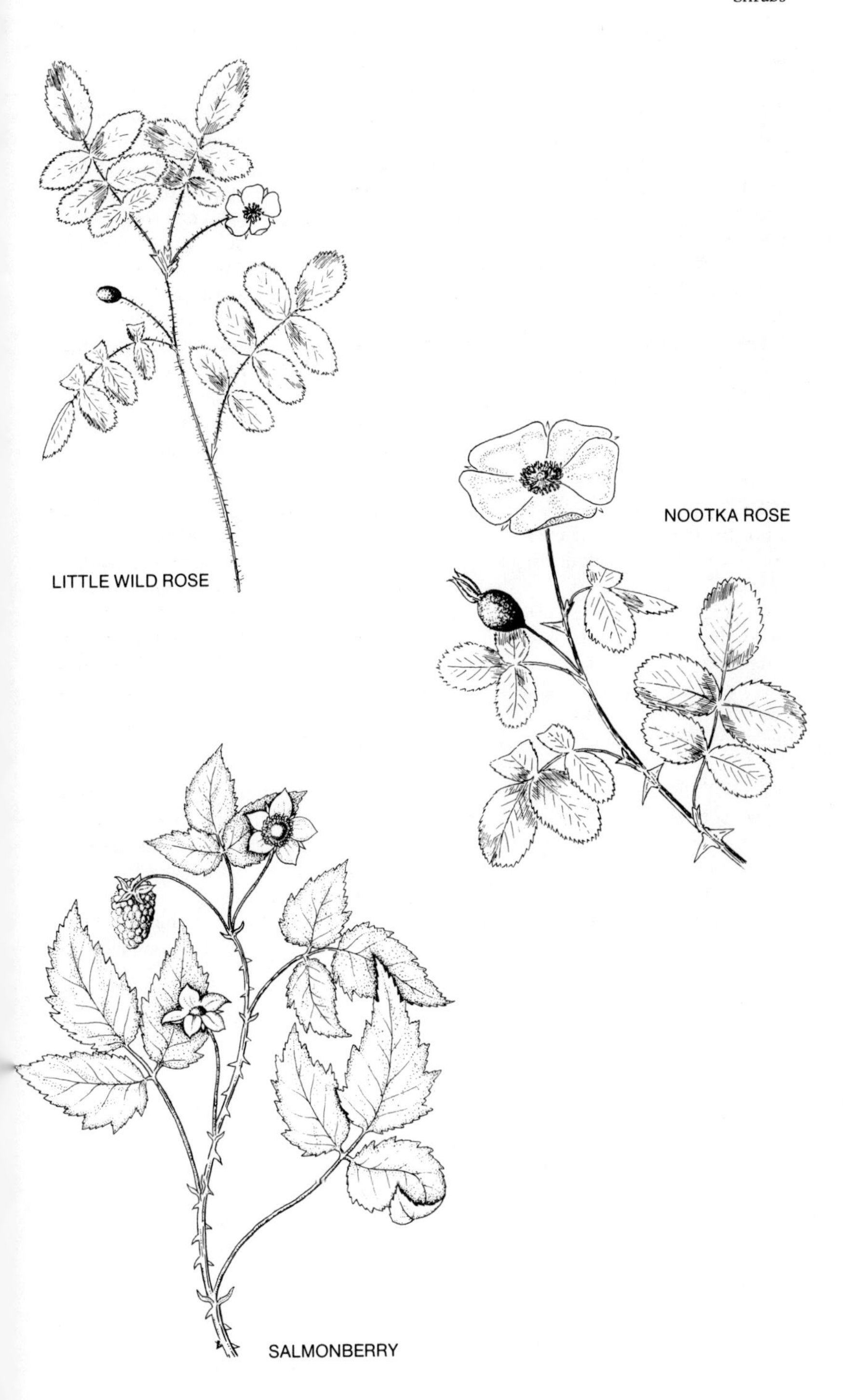
LITTLE WILD ROSE
NOOTKA ROSE
SALMONBERRY

Brown

SOOPOLALLIE *(Shepherdia canadensis)*. Soop (soap) and ollalie (berry) are the morphological roots of the distinctive name, given to this shrub by the Chinook Indians, who noted that the red fruits, when beaten in water, produce a soapy froth. A peculiar-tasting, aromatic drink was made by them in just such a manner. One of a number of species that occur chiefly east of the Cascades, this shrub is restricted in Puget Sound country to the rain-shadow areas, i.e., the northeast Olympic Peninsula, southern Vancouver and the Gulf Islands, and, of course, the San Juans, where it is fairly common. Distinctly coppery haired on the stems and leaves, Soopolallie is unlike any other shrub in our area. The leaves are dark green above and intensely white-hairy below, the margins usually curled downward. Grows 3–4 feet high with us, preferring sunbaked, gravelly forest margins; apparently most common on Lopez and Shaw Islands.

TREES

WESTERN RED-CEDAR *(Thuja plicata)*. What Washington resident is not familiar with this tree? Certainly everyone has at least heard of it, and most have seen and admired it as well. Easily distinguished by the scalelike, drooping branches and reddish, aromatic bark that tears in strips. Since long ago, cedar has been highly prized: Northwest Indians carved elaborate totem poles and other things from it, including sturdy dugout canoes and baskets. Today, cedar-chip landscaping, house designs using cedar shingles or shakes, and many other uses are common. The key to cedar's usefulness is its resistance to decay, due to a natural fungicide in the heartwood. The current high market value of cedar is reflected in the increase of "cedar rustling," now a major problem as thousands of trees are cut illegally each year.

GRAND FIR *(Abies grandis)*. From a distance, the neat symmetry and dome-shaped crown of this common San Juan tree are distinctive. Up close, the long, glossy needles, arranged in flat horizontal rows on each side of the twig, are attractive; each needle is blunt tipped, flat, and dark green. Grand fir occupies a wide ecological range in the San Juans, growing in deep, shady woods to dry, open places. Trees often grow very rapidly in full sunlight, sometimes gaining over 12 inches a year. At several San Juan localities, trees 100 feet or taller may be found, e.g., at Point Colville, Lopez Island, and near the Friday Harbor Oceanographic Laboratories.

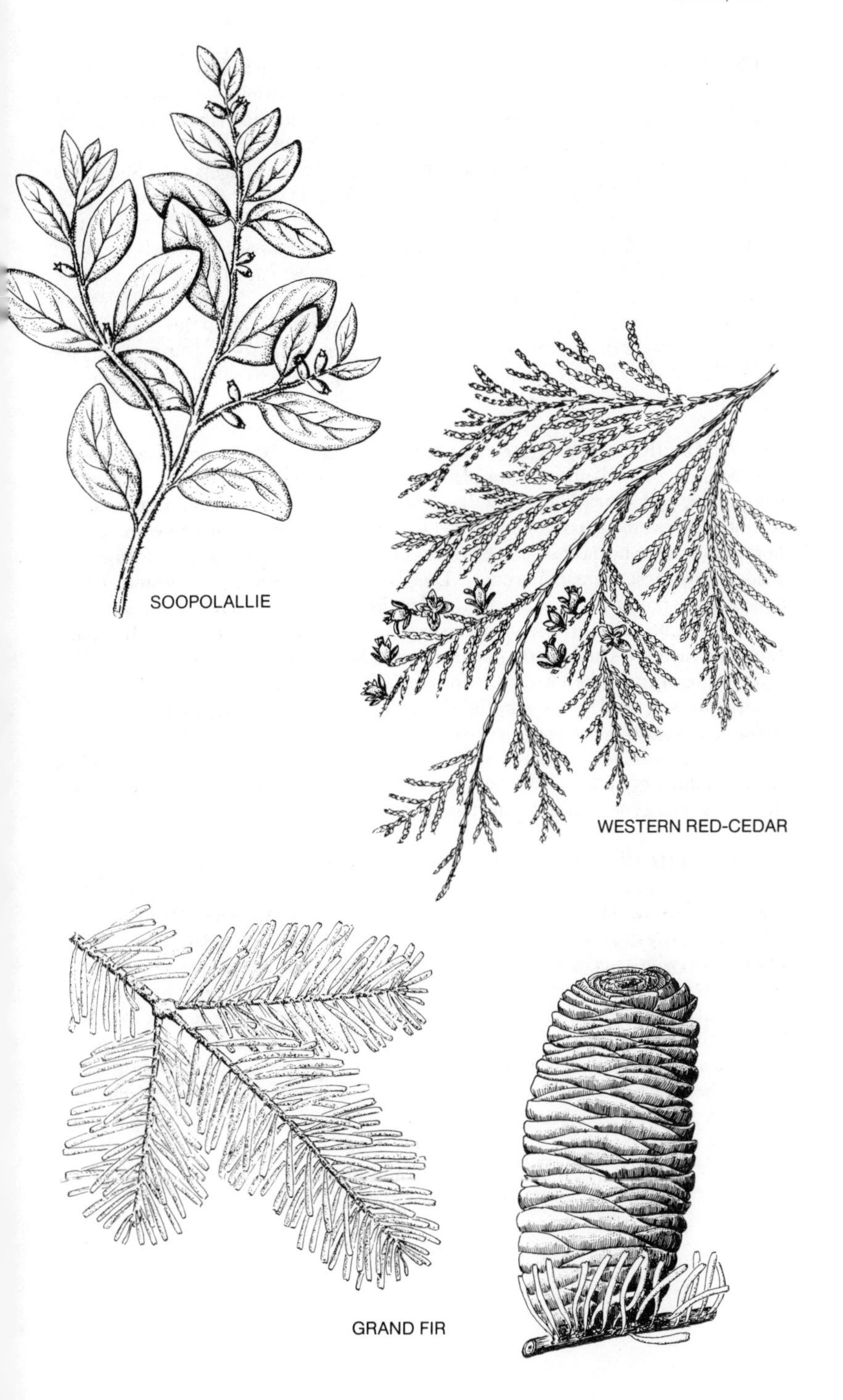
SOOPOLALLIE
WESTERN RED-CEDAR
GRAND FIR

LODGEPOLE PINE *(Pinus contorta* var. *latifolia)*. Lewis and Clark gave this tree its common name, based on their observation that the thin trunks were used widely by Indians as a support for teepees and longhouses. Recently, the name Shore pine has become prevalent, owing to the tree's propensity to form dense stands along outer coast sand dunes and other maritime habitats. Actually the Lodgepole or Shore pine occupies many habitats, including boggy lake edges, open, sandy places, rocky coastlines, and open woodlands. Seldomly growing over 60 feet in height here, it may be distinguished by the yellow-green to dark green paired needles, which average 2 inches long, and by the prickly little cones. The growth form is highly variable, depending on the site. Dense stands may be found at several San Juan localities, and solitary individuals are also common. The well-documented ability of this tree to reseed after forest fires is due to the resin-sheaths that protect the seeds. During a fire, the resin-sheath is sacrificed, burning away and allowing the seeds to fall into the ash and take root before other species do.

DOUGLAS-FIR *(Pseudotsuga menziesii* var. *menziesii)*. Washington's best-known tree, occurring abundantly in many places. This, the nation's number-one lumber tree, has provided jobs for millions and has attained world-wide distinction: it is used to reseed in deforested parts of Europe and is planted as an ornamental elsewhere. Any northwesterner who has had a cut Christmas tree should know the scented foliage well. Aside from these human purposes, Douglas-fir is of infinite importance to the ecosystems it produces, providing food for myriad creatures, from mushrooms that live on the dead branches to Brown creepers, small brown birds that scale the trunks searching for insects in the corky, furrowed bark. Mammoth Douglas-firs over 250 feet high can be seen in the Olympic rain forest. Told from other trees by the flexible needles that surround the twig, by the distinctly three-pronged bracts of the cones, and the straight growth form. A pioneer tree, it is often the first conifer to colonize an open site that has been logged, cleared, or burned. Some people may recognize the specific name that honors Archibald Menzies, the English botanist who visited Puget Sound in the 1790s. The genus name means false hemlock.

WESTERN HEMLOCK *(Tsuga heterophylla)*. A tree of the cool, north-facing slopes and valleys, with the densest stands in our area being found on Orcas and Blakely Islands. This is the most successful conifer to grow in the heavy shade of the coniferous forest floor; it will gradually come to dominate the woodland, replacing species like Douglas-fir, which needs ample sunlight. If the woodland is left undisturbed, Western hemlock will eventually form a climax forest, regenerating itself timelessly, as it does in the coastal old-growth rain forests. Thus, hemlock is the premier climax tree of most of the coastal forests from northern California to southeast Alaska. However, although supreme in deep shade growth, it is a slow-growing species, susceptible to wind damage because of its shallow rooting, and susceptible to fire damage by virtue of its thin bark. Most people recognize Western hemlock by the drooping crown and branches, but the small (½–¾ inch) needles, which are flat, blunt tipped, and often paired, are also distinctive. The cones are small and papery. Hemlock is an important source of pulpwood and tannin, the latter used in leather production.

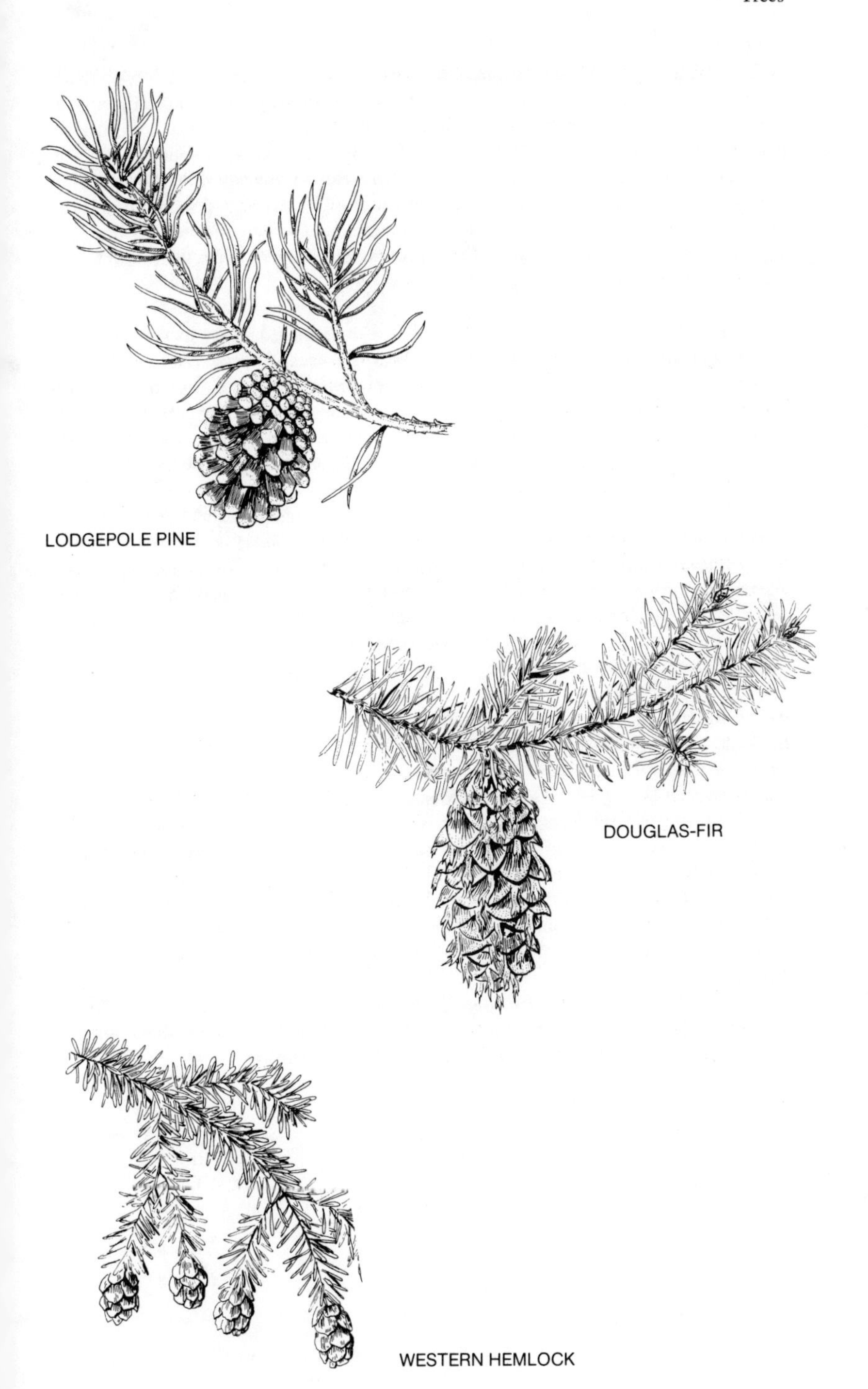
LODGEPOLE PINE
DOUGLAS-FIR
WESTERN HEMLOCK

SCOULER'S WILLOW *(Salix scouleriana).* In early spring, the pearly, yellowish white catkins appear, bringing the first sign of life to seemingly lifeless branches and, with this, the promise of warmer, sunnier days ahead. By virtue of the softness of the catkins, people often call this tree "Pussy willow." Collecting the catkin-laden boughs in bouquets is a popular early spring pastime. When the catkins fall off in April or May, Scouler's willow may still be distinguished by the obovate leaves, which are smooth and greenish above and reddish-hairy below. May reach 30–40 feet in height. Both the specific and common names honor Dr. John Scouler, a famous Northwest botanist of the nineteenth century.

RED ALDER *(Alnus rubra).* An abundant tree in disturbed, open, or swampy sites, nearly as common in the San Juans as elsewhere west of the Cascades. Persons unfamiliar with it may mistake it for a birch, because the trunks are often covered with lichens in color patterns that suggest the mottled trunks of birch. Red alder is told by its ovate, roundly toothed leaves, which are light green above in spring, dark green above in late summer, and grayish light-hairy below. The pendant, reddish yellow catkins precede the first young leaves of late March or early April. The cones are rounded and dark brown; they mature later in the season. Alder brings nitrogen to soils, making possible colonization by other species. Sitka alder *(A. sinuata)* is a shrubby species, featuring a multiple series of bent trunks, and rarely reaches 30 feet in height. The leaves are one–two times serrate, sharply toothed, and often shiny-glandular above. Scattered in the San Juans, with colonies at Spencer Spit State Park and on Mt. Constitution; small populations are also present on Cypress and Fidalgo Islands (Washington Park) just to the east.

BITTER CHERRY *(Prunus emarginata* var. *mollis).* To be expected at many wooded habitats, preferring margins but also showing up at open, scrubby places and deep woods. Variable in height and form, depending on the site: deep-woods trees are thin and may reach 40–50 feet or so; exposed-site specimens are often shrubby and 10 feet high. Note the oval, finely toothed leaves, which are usually 1–4 inches long; in fall they turn various shades of light yellowish, orange, or light red. White flowers are borne in compact clusters during spring, followed by the red, bitter-tasting cherries, which are much smaller than cultivated ones. Straight, upright branches and the grayish, black-banded bark are good field marks when flowers and leaves are absent; the notably thin twigs are also distinctive. Cherry wood decays quickly, but the bark was used by Northwest Indians in basketry. Common chokecherry *(P. virginiana* var. *demissa)* is also widespread in the islands on open rocky slopes and dry forest margins. The leaves are longer, darker green, serrate, and sharply pointed. It also has a crooked, shrubby form, black fruits, and elongate white flower clusters in spring.

SCOULER'S WILLOW

RED ALDER

BITTER CHERRY

DOUGLAS' MAPLE *(Acer glabrum* var. *douglasii).* Brilliantly displaying its scarlet-crimson leaves in the autumn sunshine, Douglas' maple adds a rare radiance to certain rocky slopes and wooded valleys in the San Juans. One is reminded of Vine maple *(A. circinatum),* well-known in most areas west of the Cascades but strangely absent in the islands except where introduced. Although the fall foliage of both trees is most celebrated, the April leaves of Douglas' maple are of an amenable fresh green color. An unobtrusive tree of the understory or edge, it seldom grows over 30 feet in height, although the islands are endowed with a few larger individuals.

BIGLEAF MAPLE *(Acer macrophyllum).* Winter's gray chill can never be far off when the large leaves of this tree begin blanketing the ground. Some of us recall having marvelled at the interesting patterns of yellow, brown, and green present in some of the leaves, or we remember as children scrambling and tumbling joyously into a neatly raked pile of them. On other occasions, we might recall our curious surprise at having discovered that some brown fungi were camouflaged in with the leaves we were raking, or we were nearly startled at the pure yellow stands of low trees on a overcast, damp day in late October. Every Puget Sound resident should know this tree, as scarcely any lowland forest lacks it. Finding old monarchs to 100 feet in height is not unusual in more mature forests, the fat, grayish trunks and branches usually bearing a beard of moss and lichens. The fine-grained wood is highly prized in furniture production and finish carpentry.

PACIFIC MADRONA *(Arbutus menziesii).* Rufous-red, bending, and curvy, the unmistakable trunks of madrona seem as if they would have been more likely sculpted by the winds of some warmer clime. Cool Mediterranean, a term used to describe the mild, sunny climate of Puget Sound's rain-shadow region, seems appropriate upon observation of this tree. Supplementing these images are the glossy leaves, which shine with health on sunny days in mid-winter and summer alike. However, it is also important to acknowledge the tree's less amiable traits, like the messy, slippery bark peelings that collect below; along with the fallen leaves, they can be hazardous, readily putting the incautious wayfarer on his back. The leaf-bark litter also tends to stifle understory vegetation and, in fact, pure stands of young madrona present at several San Juan locations are notable in this regard, i.e., the understory is absent. And even though a person may be attracted by its pleasant foliage, madrona is also virtually impossible to transplant successfully. Nonetheless, one need not overstate the case; the tree's beauty is adequate compensation. The bright reddish berry clusters appear in late summer and last until late fall; scores of Varied thrushes gulp them down and, if one is watchful, a nervous Hermit thrush may be found among them. Preferring hard-packed soils or cracked bedrock for growth, Pacific madrona is especially numerous in the San Juans, sometimes reaching 80 feet in height. Finally, a peculiar feature is the coolness of the bark, which remains so even on the hottest days.

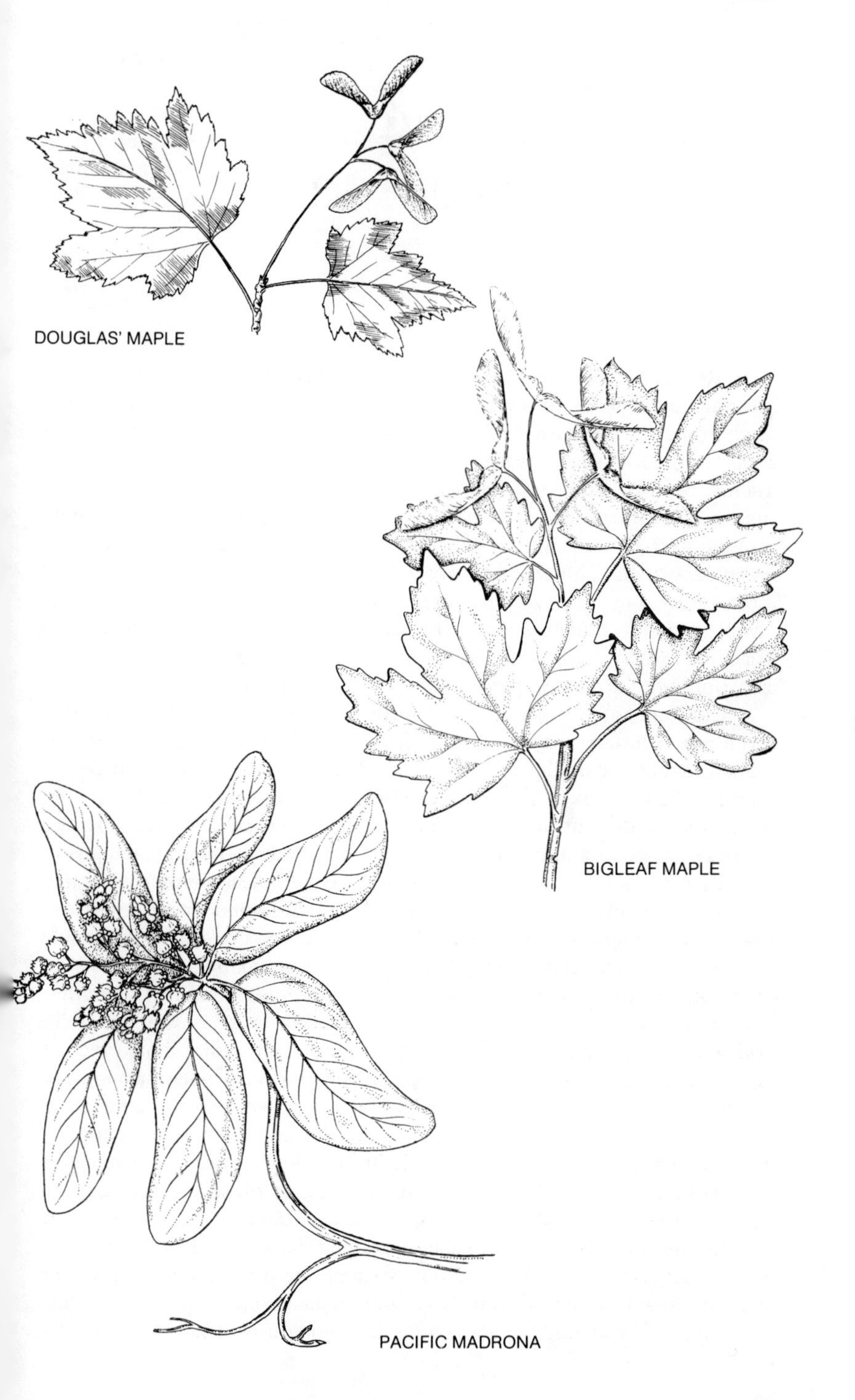
DOUGLAS' MAPLE
BIGLEAF MAPLE
PACIFIC MADRONA

4
MARITIME

Relentless wave action and the impact of glaciation are the two most prominent influences on San Juan coastlines. Waves gradually carve seemingly impenetrable rock walls and reduce steep, glacially deposited gravel banks to gentle, sandy beaches. The extremity of nature is the dominant theme. Shifting sandbars, dry, windswept soils, high salinity, and intense summer sunlight—these are characteristic of many coastline environments, conditions that would quickly kill many plants found in adjacent uplands and protected woods. Only the rugged coastline plants, with their special adaptive protections, can colonize this margin between two worlds. The seaside dwellers are much in evidence in the San Juans because of the many miles of coastline. Rocky headlands, sandy coves, and salt marshes are typical, constituting a substantial percentage of the total habitat.

Uplift of land masses, caused by plate tectonic activity and other factors, led to the rocky character of the San Juans. Rocky, steep shorelines are very numerous and, especially on many of the smaller islands, are the only type of maritime habitat one is likely to encounter. Sandy or cobble beaches are less frequently found but are prominent on Lopez and San Juan Islands. Salt marshes are yet less common but are well represented at several locations on the larger islands. Of the three habitats, the last two clearly have the greatest number of species unique to the maritime habitat. All three vary year to year in species composition, but the shady beach seems to be particularly changeable.

Seaside flora adapts to a variety of environmental constraints, as mentioned earlier. Open sites are subject to intense sunlight in summer, causing dehydration. Heat retention by sandy soils causes severe heat to be inflicted on those species present. Indeed, one need only walk a short stretch of sandy beach during a hot July day to gain an appreciation of what the roots and foliage of the plants must endure. Sands hold little moisture or nutrients, being quickly leached by rainfall. Strong winds and shifting sandbars at some sites necessitate sturdy root systems. The net effect of these stresses is the tendency of maritime species to be lowgrowing, matted, or prostrate, spreading by extensive root systems, rhizomes, or aerial roots. In addition, many species are succulentlike or have hairy or mealy-surfaced foliage to insulate themselves from overheating and dehydration. Several maritime dwellers grow in positions that minimize the impact of sunlight on their foliage, i.e., with their leaves placed at an angle, not allowing the sunlight to strike them directly.

Several species have grayish leaves (e.g., members of the Goosefoot family) that absorb less sunlight.

Common orache *(Atriplex patula)* typically grows on sandy or low rocky beach situations, usually just above the high tide mark. Pickleweed or Salicornia *(Salicornia virginica),* the dominant element of most Puget Sound salt marshes, is immersed by incoming tides. On sandy beaches, Yellow sand-verbena *(Abronia latifolia)* displays compact heads of yellow trumpet flowers throughout spring and summer. Impressive colonies of this species (and of several other halophytes) are present at South Beach, San Juan Island. Tall peppergrass *(Lepidium virginicum)* grows scattered among other seaside flora on sandy or pebble-strewn beaches and sometimes on adjacent rocky bluffs. No substantial sandy beach is without the creeping stems of Japanese beach pea *(Lathyrus japonicus)* and the dark green, sprawling mats of Silver bursage *(Ambrosia chamissonis).* Perhaps most abundant of all maritime species is Puget Sound gumweed *(Grindelia integrifolia).* Its showy yellow flowers often persist into late autumn. Pacific cinquefoil *(Potentilla pacifica)* is locally abundant at certain salt marshes, less common at sandy beaches.

Although more commonly associated with rocky soils or meadow environments, several more species seem to grow exclusively within sight of salt water. Rocky Mountain juniper *(Juniperus scopulorum)* is one such species, growing on rocky bluffs adjacent to salt water. Lance-leaved stonecrop *(Sedum lanceolatum)* and Sea thrift *(Armeria maritima)* are of the same habitat. Naked desert parsley *(Lomatium nudicaule)* is often present on less disturbed meadows near the sea.

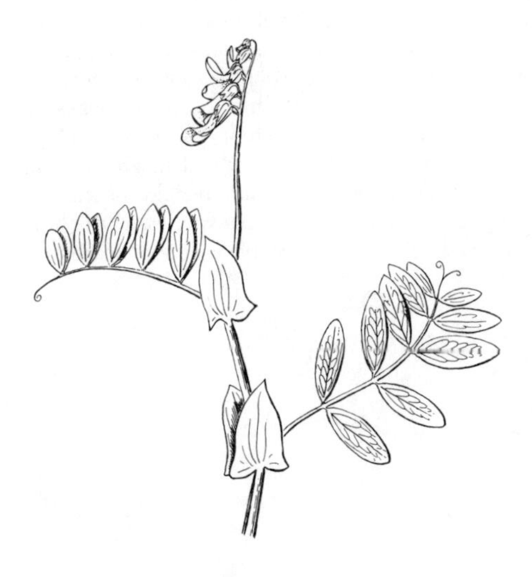

GRASSES AND GRASSLIKE PLANTS

SEASIDE ARROW-GRASS *(Triglochin maritimum).* First collected in our area by Lyall, who found it on Lopez Island in 1858, Seaside arrow-grass is common throughout the San Juans on salt marshes. Spencer Spit State Park, Lopez Island, is an excellent place to see it. Particularly during May and June, the small greenish blooms, often strikingly tinged with red or white, are conspicuous, scattered along the tall stems. The solid stems themselves rise to several feet in height, remaining in dried form long after blossoming. The basal rosette of narrow, stiff leaves, dwarfed by the stem, is highly suggestive of Sea plantain. The stems of this distinctive species sometimes grow in a crooked and erratic manner to 3 feet or more.

FLOWERS

White

TALL PEPPERGRASS *(Lepidium virginicum* var. *menziesii).* During late spring and summer, tiny white flowers are borne at the tips of the stems of this species. At other times, note the rotund, flattened seed pods (silicles). The basal leaves are pinnate and crisp-hairy; stem leaves tend to be smooth and linear. To 12 inches tall. In the San Juans, it is frequently found on sandy or rocky beaches and adjacent meadows.

PACIFIC WATER-PARSLEY *(Conioselinum pacificum).* The sandy or rocky seashores of shady, protected coves seem to be the preferred habitat of this perennial. The leaves are fernlike, smooth, and dark green, fading to yellowish. The tiny white flowers are borne in umbels during June–July. Usually short stemmed, but occasionally several feet tall. Common and widespread in the San Juans. May be confused with Poison-hemlock *(Conium maculatum)*, which is much taller, has larger, lacy leaves, a purple-spotted or streaked stem, and occurs commonly on roadsides in our area. Deadly poisonous.

SALT MARSH DODDER *(Cuscuta salina).* Have you ever walked through a salicornia salt marsh and noticed a small area covered with vivid orange? Closer observation of this area will reveal a wiry substance that seems to be of a most unplantlike texture, reminding one more of plastic or silly string. This peculiar creeper is Salt marsh dodder, a species whose chief means of subsistence is parasitizing salicornia. This is accomplished by sending root probes, called haustoria, into the stems of its host. As it does not need leaves or chlorophyll, no green pigment is present. The small whitish flowers, which appear in late summer, do not seem to be very important for the plant's propagation. An unmistakable species, as the bright orange color is readily seen from a fair distance; occasionally attacks seashore plants other than salicornia.

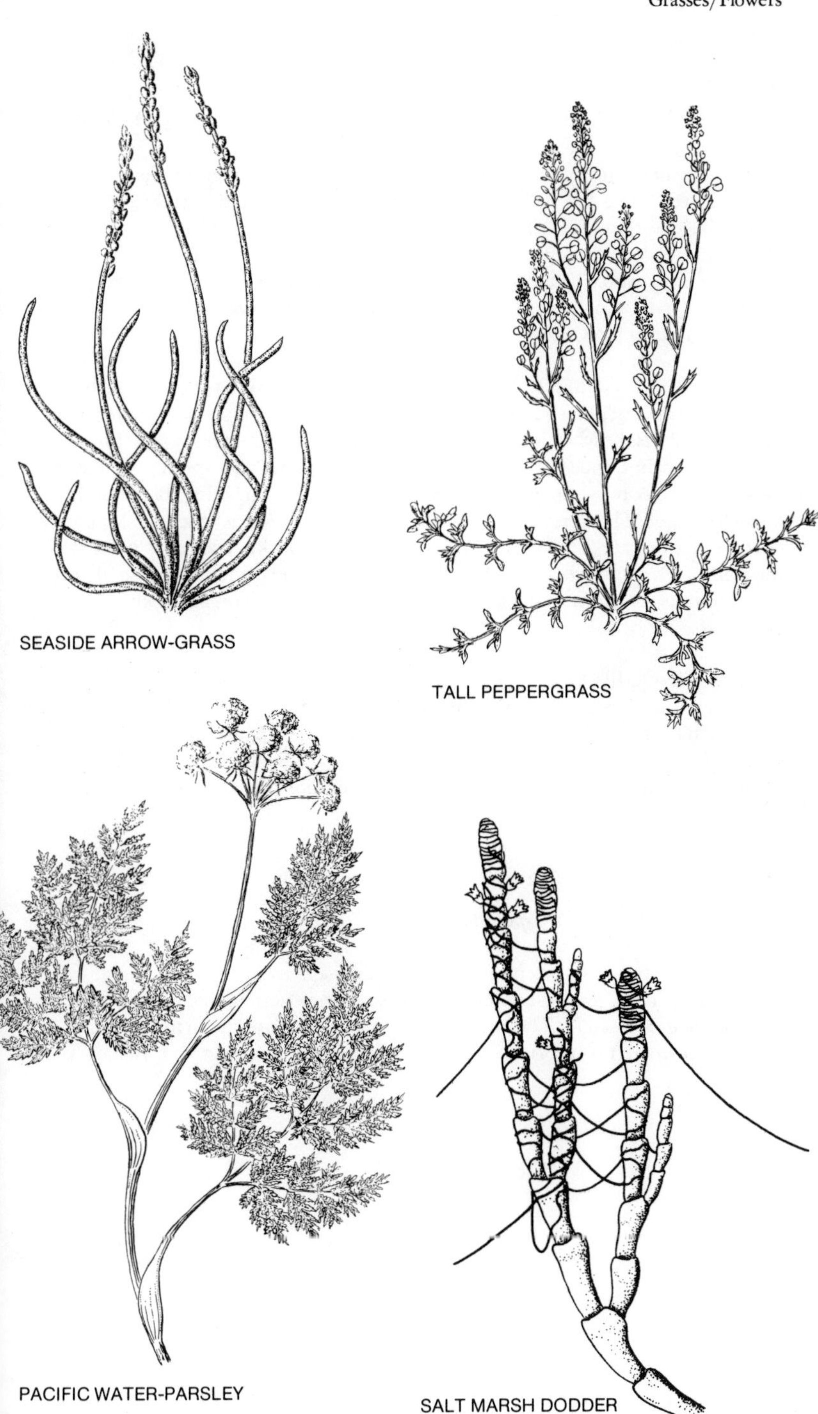

SEASIDE ARROW-GRASS

TALL PEPPERGRASS

PACIFIC WATER-PARSLEY

SALT MARSH DODDER

Yellow

YELLOW SAND-VERBENA *(Abronia latifolia)*. The exquisitely fragrant, brilliant yellow flowers of this beach resident will be forever remembered by us in association with the cry of circling gulls on a breezy summer afternoon at South Beach, San Juan Island. Here, where the most extensive colony of Yellow sand-verbena in the area occurs, one may observe easily the neat symmetry of the flowers. These flowers consist of a cluster of long trumpets that closely hugs the sand. Dark green leaves—thick, succulent, and covered by hairs—may be found throughout the area. Most abundant on San Juan Island, but occurs locally elsewhere. An exclusively coastal species that is at its northern limit in the Queen Charlotte Islands.

PUGET SOUND GUMWEED *(Grindelia integrifolia* var. *macrophylla)*. Perhaps the most abundant maritime wildflower in our area, found in any seaside environment—on coastal bluffs, beaches, salt marshes, and meadows. The pure yellow flowers are large (2–3 inches wide) and showy atop the tall stems and are among the most persistent of bloomers, appearing in June and sometimes flowering, albeit rarely, well into December. The spreading involucre bracts have inspired the name gumweed, because they excrete a milky latex and are consistently sticky. The genus name honors the eighteenth century botanist, David Grindel. Occasionally found in a smaller form away from its maritime haunts, as on some rocky meadows on Mt. Constitution and elsewhere.

PACIFIC CINQUEFOIL *(Potentilla pacifica)*. Common along swampy lake margins as well as salt marshes and other saline habitats throughout our area. During June and July, watch for the yellow flowers, reminiscent of buttercup, that are borne on leafless stems. More obvious most of the time are the strongly divided leaves, which are shiny green above and white-hairy below; in ideal conditions they may be quite large, to 6 inches or more long.

Pink

BEACH SANDSPURRY *(Spergularia macrotheca)*. Salt marshes, gravelly meadows adjacent to the sea, the rocky coastal bluffs are home to this, our most common and showiest sandspurry. Its luscious pink "stars" blossom in July–August, but at least as important in identification are the thick, procumbent stems, which are sticky-haired like the leaves and sepals. The largest of four *Spergularia* species in the San Juans, it is widespread; an excellent place to encounter it is Eagle Cove, San Juan Island.

YELLOW SAND-VERBENA
PUGET SOUND GUMWEED
PACIFIC CINQUEFOIL
BEACH SANDSPURRY

Purple

JAPANESE BEACH PEA *(Lathyrus japonicus).* Creeping across an area as only the maritime sprawlers can, the long stems of this pea are a predictable sight at any major sandy beach. The leathery, fleshy leaflets, usually in groups of six–twelve, are resistant to water loss. Below, strong roots prevent wind uprooting. From April to July, two–eight purplish flowers appear, each with whitish wings and keels. One of several peas in our area, this is the most abundant one at maritime sites. Found from Alaska to California, and in the Far East as well.

Blue

SEASHORE LUPINE *(Lupinus littoralis).* Cattle Point, San Juan Island, is the easiest place to find this lupine, as it dominates the sandy areas near the lighthouse. The prostrate stems, to 3 feet long or more, bear symmetric whorls of five–ten leaves each and, during May to August, bear the purplish blue white-flecked flowers in the terminal, low spikes. The name *littoralis* means of the beach, and in fact, this is a chiefly maritime species, occurring in scattered patches in the San Juans, particularly on Lopez and San Juan Islands.

Green

COMMON ORACHE *(Atriplex patula* ssp. *hastata).* Like many of the Goosefoot family, an excellent edible species, especially the young leaves. Mature leaves are highly variable in shape, from nearly ovate to linear; the fleshy texture, grayish green color, and mealy surfaces are characteristic. Orache is apparently an old Arabic name; the Latin species name refers to the wide-spreading growth form, as the up to 2-foot stems often branch. New Zealand spinach, widely cultivated in our area, has a similar taste, being a member of the Goosefoot family also. Common orache is abundant along beaches throughout the islands.

SALICORNIA *(Salicornia virginica).* The dominant species of salt marshes, covering muddy, protected bays with broad mats. In adapting to daily tidal emersion and unstable rooting conditions, Salicornia has developed a tough root system and spreads mainly by creeping. Correspondingly, the need for flowers and leaves is diminished, and as a result, both are present but in much reduced form, i.e., minute and scalelike. The segmented growth is fleshy, edible, and salty to the taste, and is an important food source for many creatures, including geese and other waterfowl. Large salicornia salt marshes can be found at Fisherman Bay, Lopez Island, at Argyle Lagoon, San Juan Island, and at many other places.

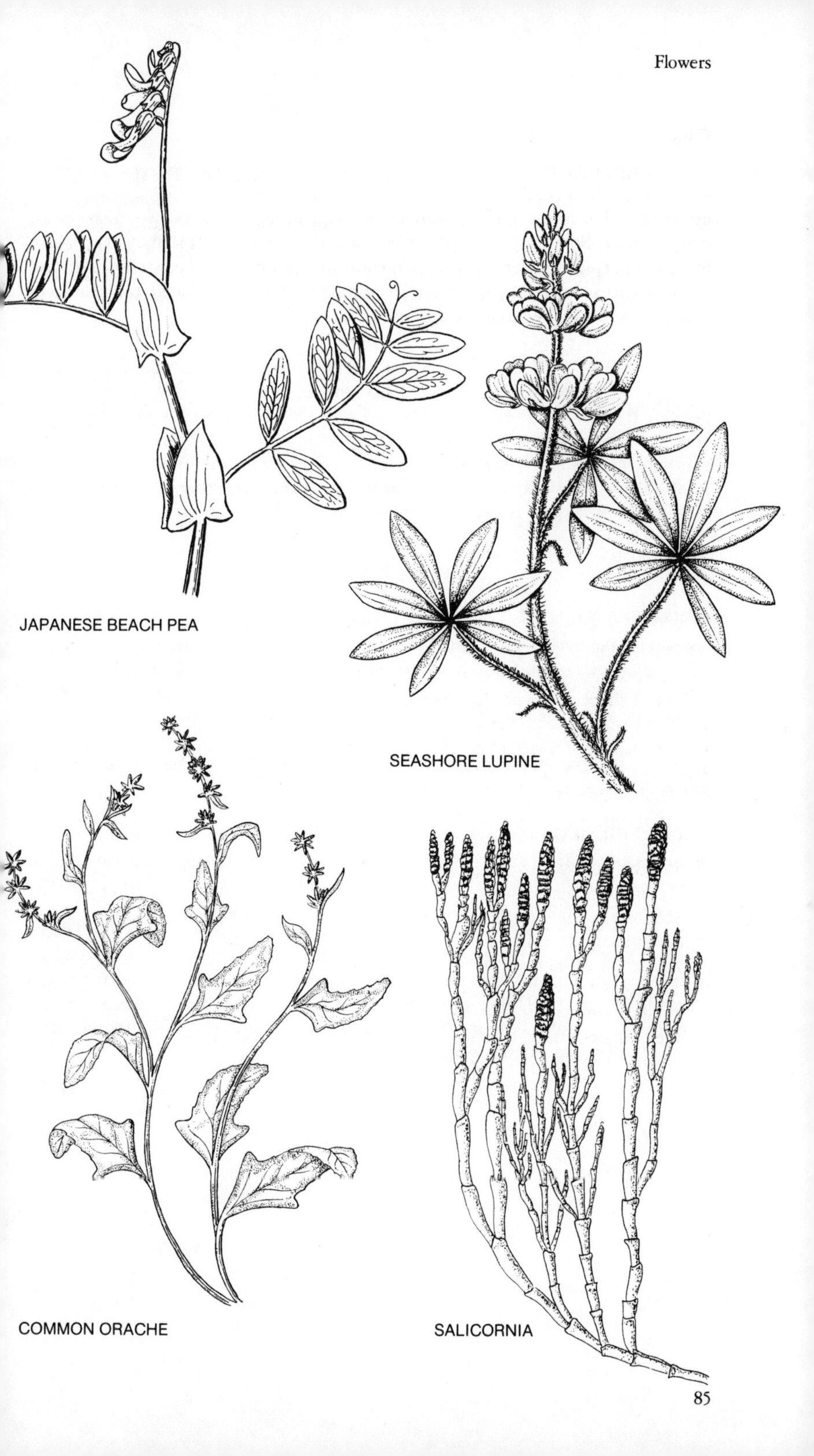
JAPANESE BEACH PEA
SEASHORE LUPINE
COMMON ORACHE
SALICORNIA

Green (continued)

SILVER BURSAGE *(Ambrosia chamissonis* var. *bipinnatisecta).* The sprawling, dark green foliage of this sand lover is familiar throughout Puget Sound country, with the San Juans no exception. The strongly divided, pinnate leaves are irregular in shape and are the most obvious feature. During summer, spikes bear both fertile and sterile flowers, the latter on top and the former below, extending into the leaf axils. These greenish blooms have small, burlike involucres (see glossary), hence another name, burweed. A large species with stems often over 5 feet long.

SEA PLANTAIN *(Plantago maritima* var. *juncoides).* Along rocky bluffs exposed to salt spray, sandy or pebble-strewn beaches, and salt marshes, this "leatherneck" of a plantain thrives. The ascending leaves, dark green, thick, and pointing straight up, are very distinctive, arranged in a neat basal rosette. Throughout much of the growing season, a leafless spike (scape) is evident at the center of the rosette, bearing many small, greenish flowers. In its various forms, Sea plantain occurs on the coasts of much of North America and Eurasia. Common in the San Juans.

TREES

HOOKER'S WILLOW *(Salix hookeriana).* Growing along woodlands, scrubby areas, and sand dunes in close proximity to salt water, this is the common willow of the coast. Its range, which is limited to Southern British Columbia to northern California, lies strictly west of the Cascades. Similar to Scouler's willow *(S. scouleriana)* of the woodlands section, but differing from it in having wider leaves that are hairy on both surfaces. Named for Joseph Hooker (1817-1911) who, like Douglas, covered a lot of ground in search of new flora and, as a result, has a number of species named for him. Hooker's willow is fairly common in the San Juans.

PACIFIC CRABAPPLE *(Pyrus fusca).* Often somewhat obscured by its interwoven branches, the straight spurs of Pacific crabapple have torn the clothing of more than one unwary traveller who has ventured through its thickets. A variable species in terms of foliage, form, and habitat. Often abundant near swampy places, open meadows, and seashores, where it typically grows dense and shrubby. Also found at woodland edges, where it may be a slim understory tree 30–40 feet in height. Though the leaves are usually toothed and trilobed, they are often irregularly shaped and thus may not be instantly recognized. Unlike those of most of our trees and shrubs, the leaves turn attractive shades of yellow and light red in fall. The white blossoms, present during April and May, are pleasingly fragrant.

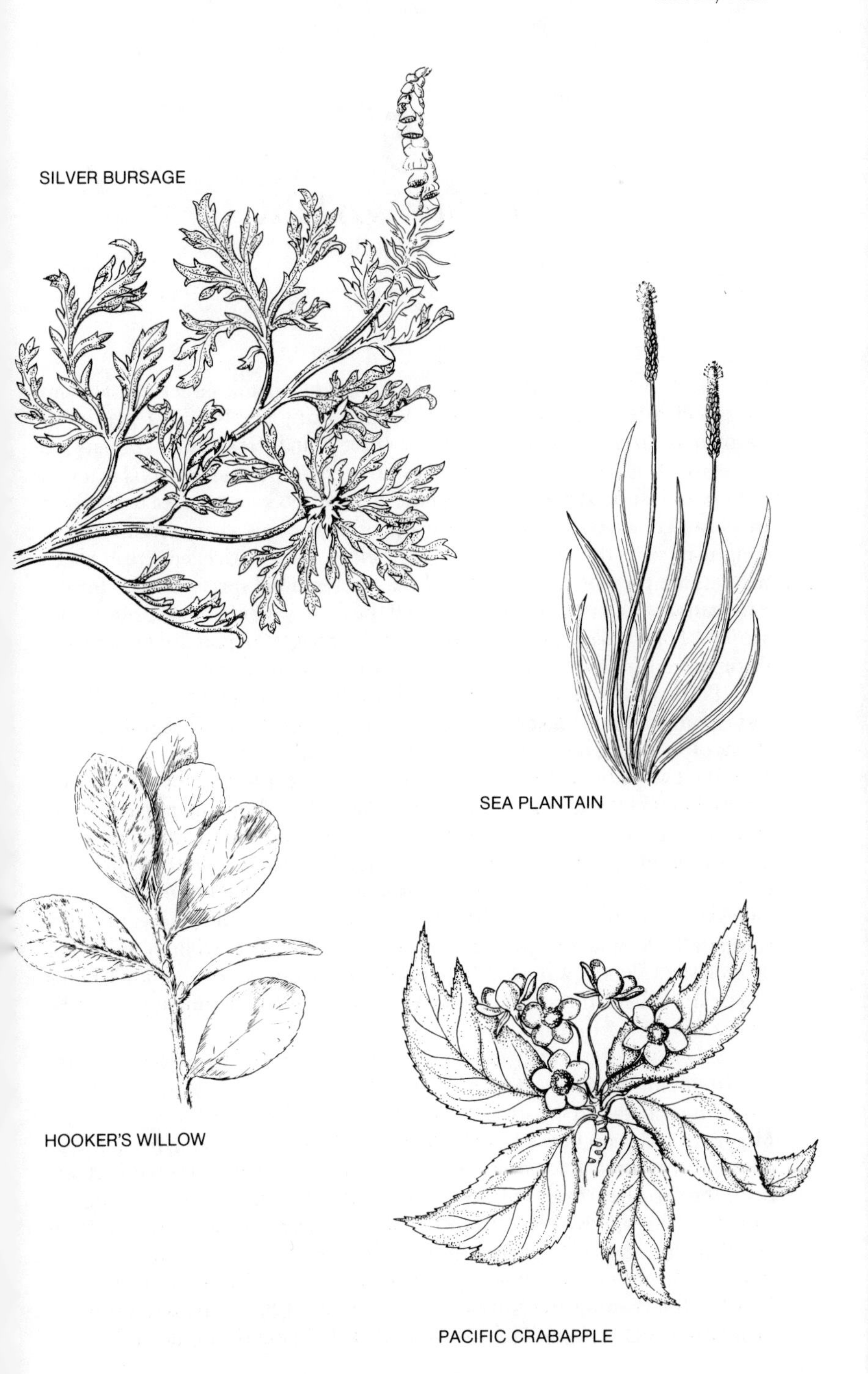
SILVER BURSAGE
SEA PLANTAIN
HOOKER'S WILLOW
PACIFIC CRABAPPLE

5
FRESH WATER

Fresh water habitats are most prominent on the larger islands and consist of a few lakes of considerable size, many ponds, and a paucity of permanent streams. However, even on these islands fresh water habitats are far less in evidence than salt water ones, and certainly less understood in terms of flora and fauna. Actually, much of the archipelago's surface area is devoid of fresh water, especially on the smaller islands. This characteristic seems congruent with the dry climate. Nonetheless, several excellent fresh water habitats are present. In fact, some of the most unusual species known for the San Juans are of the fresh water zone. Of the fresh water habitats present, the shallow, boggy lake type, particularly where sphagnum mats are present and disturbance is minimal, is by far the most floristically diverse.

Permanent streams are scarce in the San Juans, occurring chiefly on Orcas, San Juan, and Blakely Islands. Aside from their general insignificance in terms of numbers, they are of little note botanically for several reasons. First, with the exception of some of the streams on Orcas Island, the streams are reduced to trickles during summer and, thus, have a declining impact on the surrounding environment. Second, several of the streams are heavily shaded, as at the drainage into Cascade Lake on Orcas Island, although there are a few distinctive species here. Finally, the isolation of San Juan County's stream valleys seems to limit the spread of typical species and has kept them comparatively underdeveloped. For instance, Western trillium *(Trillium ovatum)* and Slender waterleaf *(Hydrophyllum tenuipes)*, two common herbs of stream valleys throughout most of western Washington, appear to be absent from the San Juans.

Lakes and their margins, as the largest of fresh water habitats, are correspondingly the most rich botanically. However, it is important to differentiate between the steep-sided, deep lake type, e.g., parts of Cascade and Mountain Lakes on Orcas Island, and the shallow, boggy lake type. Without question the latter is far more interesting botanically and otherwise. Aside from the fact that several of San Juan County's endangered and threatened vascular plant species come from the shallow lake habitat, such places are excellent for observing myriad insects, from the ferocious water tiger to colorful dragonflies, as well as frogs, salamanders, Wood ducks, and other life. Lakes such as Summit and Killebrew on Orcas Island, Sportsmans–Egg on San Juan Island, and Spencer–Horseshoe on Blakely Island fit this description.

Pacific willow *(Salix lasiandra)* is an integral part of any substantial riparian situation in the islands. Hardhack *(Spiraea douglasii)* is often found near lakesides and boggy places, displaying fragrant pink plumes in summer. Common monkey flower *(Mimulus guttatus)*, a species noted for its showy yellow flowers, is locally common in roadside ditches and sometimes on lake edges. Speedwells (particularly *Veronica americana* and *scutellata)* are quite common in nearly any fresh water habitat, as are various buttercups *(Ranunculus)*. It would be more appropriate to call Field mint *(Mentha arvensis)* marsh mint in our area, since that is where it occurs. Small bedstraw *(Galium trifidum)* is common at lake edges.

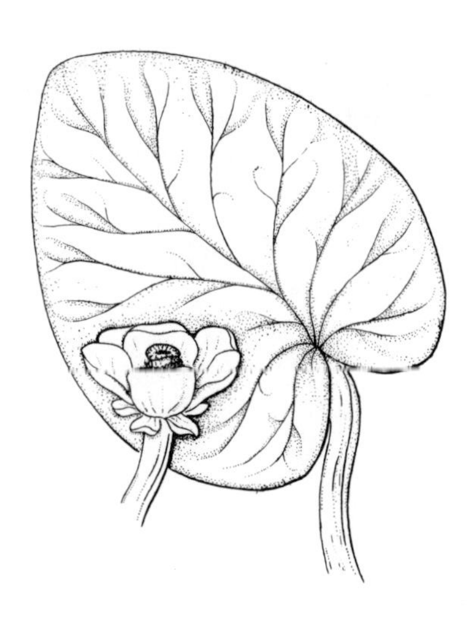

GRASSES AND GRASSLIKE PLANTS

SLOUGH SEDGE *(Carex obnupta).* This is our most abundant sedge, occurring repeatedly at poorly drained places, although many other sedges are present in the San Juans. The leaves are long, narrow, and shiny-glossy; the prominent edges are often sharp and the leaf midvein is obvious. A large sedge, sometimes 3 feet tall, with elongate, frequently nodding spikes. Spike and leaf tips typically turn brownish with age. Found from southern British Columbia to California.

COMMON MARE'S TAIL *(Hippuris vulgaris).* The dark green, erect stems and whorled leaves of this species, arising from shallow fresh water, cannot be mistaken for anything else. On first impression, the peculiar spikes may remind one of young conifers. The genus name can be split into *hippos* (horse) and *aura* (tail) from the Greek, although the species does not seem to have any particular resemblance to a horse's tail. Occurs over much of North America; locally abundant, especially on Orcas Island, where dense colonies dominate in several areas.

BROAD-LEAVED PONDWEED *(Potamogeton natans).* By far the most common pondweed in the San Juans, often growing with Indian pond-lily. Pond surfaces are often completely dominated by the masses of ovate to nearly round leaves, which tend to be more symmetric and more evenly light green than those of Water smartweed. The drab, greenish flowering spikes appear in mid-summer. Below the surface are the long, weak stems, which bear grasslike leaves completely unlike those that grow above the water. Found over most of North America and Europe.

FLOWERS

White

WATER-PARSLEY *(Oenanthe sarmentosa).* A dominant perennial of many poorly drained locations, this plant is found almost entirely west of the Cascades, from Alaska to California. It grows in deep, shady woods where small pools, streams, or muddy situations exist, as well as in more open wet places like lake edges or ditches. The weak, succulent stems branch freely in either an ascending or a matted form, often rooting at the nodes (the Latin name *sarmentosa* means "bearing runners," a reference to this characteristic). During late spring and summer, umbels of creamy white flowers rise above the matted growth, frequently giving off an unpleasant odor. Abundant in the San Juans.

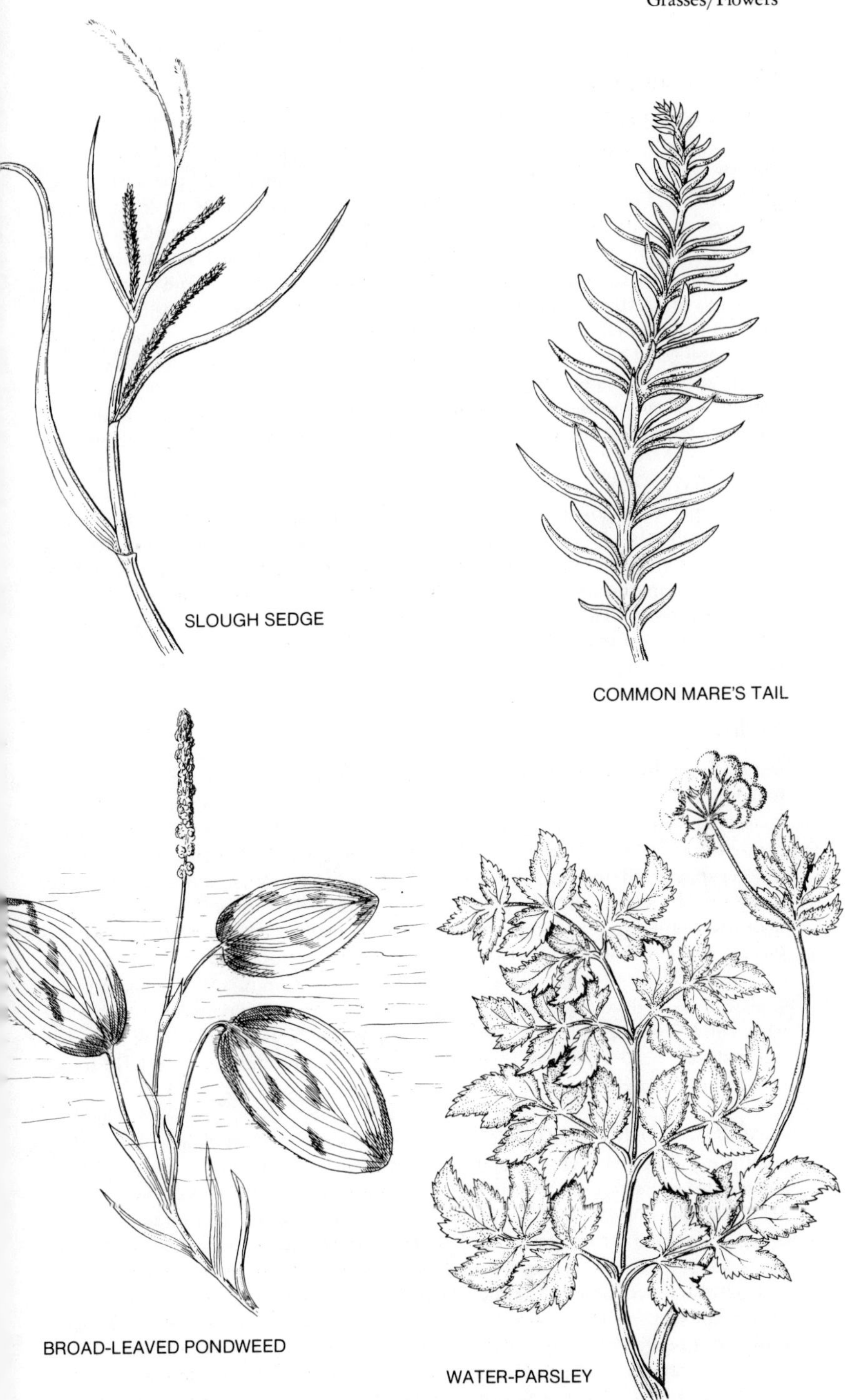
SLOUGH SEDGE
COMMON MARE'S TAIL
BROAD-LEAVED PONDWEED
WATER-PARSLEY

White (continued)

COW-PARSNIP *(Heracleum lanatum)*. A huge species that may strike the observer as being something out of the tropics rather than a native of the Pacific Coast. Appropriately named for Hercules, this wet-ground mammoth might be better named giant parsnip, since everything about it is outsized. The thick stems are often 4 inches in diameter, while the heavy, broad leaves are often 12 inches long. Although the greenish white flowers, present in May, are ill smelling, the dried clusters are used in bouquets. Plants routinely reach 6–8 feet in height, and one specimen seen several years ago in Seattle was approximately 15 feet—an incredible height! The hollow stems are substantial enough that some innovative residents on the Olympic Peninsula have employed them as makeshift irrigation piping for small gardens. Both stems and foliage are readily eaten by cattle, bears, elk, and, formerly, some of our Northwest Indians. Common in wet seeps, open stream valleys, and on coastal bluffs west of the Cascades, especially on the outer coast. Usually near salt water in the San Juans.

SMALL BEDSTRAW *(Galium trifidum)*. From one fresh water site to the next, this little creeper shows an unfailing abundance. During July and August, the small white flowers, borne in clusters of one–three at the tips of axillary branches, are fairly conspicuous. At other times, however, the small, blunt-tipped leaves (usually in whorls of four–five) are the best field mark, although they may be inconspicuous in the usual tangle of growth. The weak stems are often several feet long and readily climb on surrounding vegetation. One of several bedstraws in the area, but the only one so restricted to the fresh water habitat.

Yellow

INDIAN POND-LILY *(Nuphar polysepalum)*. Because of its size and prevalence, Indian pond-lily is a critical component of the ecology of many lakes and ponds. Tiny microorganisms feed on the growth and, in turn, become food for larger insects—damselfly larvae, water boatmen, whirling beetles, etc. Fish, salamanders, and frogs congregate around and on pond-lilies to feed on these insects, only to be devoured themselves by herons, ospreys, or larger fish. As the adult dragonflies emerge, along with damselflies and other insects, they feed on smaller aerial creatures above the surface and are themselves devoured by swallows. But aside from the pond-lily's ecological role, its roots and foliage were eaten by Indians in times of famine, and the seeds, called *wakos,* were always enjoyed as a food. Certainly everyone has seen the lily pads, but fewer may have noticed the stunning yellow flowers, which look like great teacups that have emerged from the depths. Blooms during June–July; found throughout the area.

COMMON MONKEY FLOWER *(Mimulus guttatus* var. *guttatus)*. The showy goldfinch-yellow flowers add interest to any setting. They have obviously hairy throats and reddish spots on the lower lip and are arranged in showy clusters. The stems are fleshy and tall—up to 2 feet. Although there appears to be some uncertainty as to its origin, the common name refers to either the reddish blotches, said to be shaped like monkeys, or to the whole flower, supposedly shaped like a monkey face. Common at several fresh water habitats, blooming from May to September. Two other forms occur in the San Juans. Var. *depauperatus* is a smaller form, found on drier rocky sites scattered through the islands; it is not runner-bearing, and has toothed leaves. Var. *grandis* is restricted to open seeps near salt water, especially along the southern edges of San Juan Island. It is larger—often 3 feet tall, and is smooth to weakly hairy.

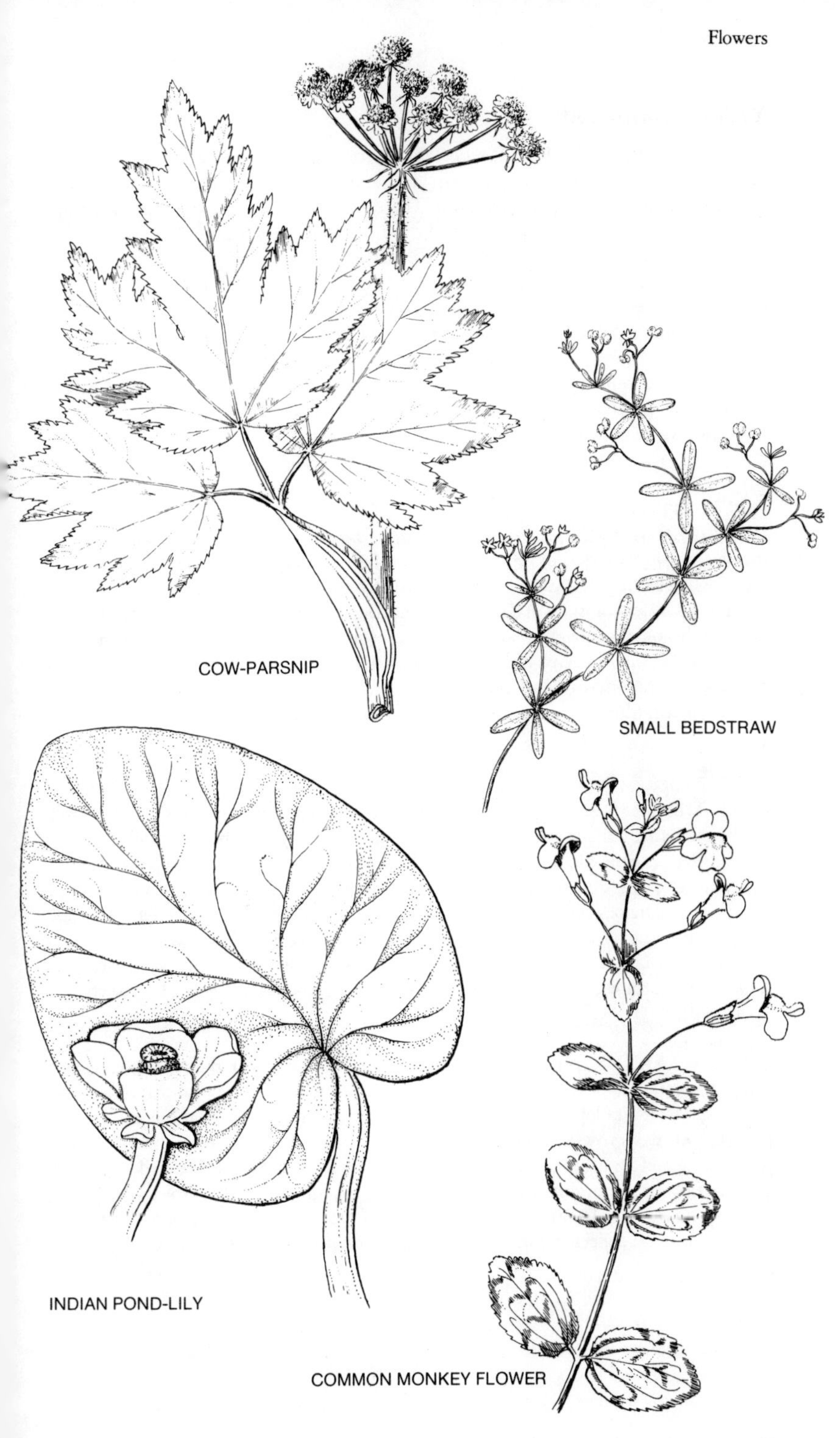
COW-PARSNIP
SMALL BEDSTRAW
INDIAN POND-LILY
COMMON MONKEY FLOWER

Yellow (continued)

CREEPING SPEARWORT *(Ranunculus flammula)*. The cheery yellow flowers of this buttercup may often be found poking through the choking growth of sedges, herbs, and grasses along a pond or lake edge. The quickest check against other buttercups (there are several in the San Juans) are the entire, elliptical (spear-shaped) leaves. Like several other buttercups and other denizens of the fresh water site, Creeping spearwort roots nodally. Blooms from May to September. The common buttercup around low fresh water areas, often in muddy situations; usually does not venture out into the water, but grows within 3–20 feet of shoreline. White water buttercup *(R. aquatilis* var. *hispidulus)* is entirely aquatic, the weak stems collapsing when removed from water. Its flowers are yellowish white and two leaf types are present—the trilobed, entire upper leaves (at the surface) and the deeply dissected, linear ones (well below). Less common, but widespread on the larger islands.

Pink

WATER SMARTWEED *(Polygonum amphibium)*. The lush pink flower heads of this curious aquatic species are conspicuous when they emerge on the surface of lakes and ponds during July–August. Below the surface are freely branching stems that are sometimes 20 feet or more in length. The Latin species name is accurate, for the plant is not wholly aquatic; upright plants occasionally grow near the water's edge. When not in bloom, the open groups of floating leaves are readily visible; they tend to be dark greenish and oblong-ovate. A species of worldwide occurrence and rather common in the San Juans, often growing near Indian pond-lily.

Purple

FIELD MINT *(Mentha arvensis)*. Because it is seldom found in fields, some people prefer the name Canada mint, although the Latin species name means of fields. Watch for the light purple blossoms, borne in clumps at the leaf axils, during July–August. Mints have square stems; Field mint's are also very hairy. Often abundant along lake or pond edges. The heavenly delicious scent of mint arises from the crushed leaves. Field mint (and other mints) contain menthol, as the Latin name *mentha* implies. Spearmint *(M. spicata)* occasionally occurs along roadsides and in wet places. Its flowers are terminal, not axilliary, and the leaves are stalkless, as opposed to the stalked Field mint.

COOLEY'S HEDGE-NETTLE *(Stachys cooleyae)*. Although the foliage suggests Stinging nettle, Cooley's hedge-nettle is in fact a mint and will not punish those who brush against it. Cooley's hedge-nettle prefers damp, often shady situations, although it does not always grow in poorly drained areas. The reddish purple flowers are 1–2 inches long and are arranged on a terminal spike or in the leaf axils. The striking beauty of these flowers, blooming from spring through fall, compensates for the coarse, hairy, ill-smelling foliage. The tuberous, starchy rootstalks are edible. Named for Dr. Grace Cooley, who first collected it in 1891, on Vancouver Island. Common and widespread in the San Juans.

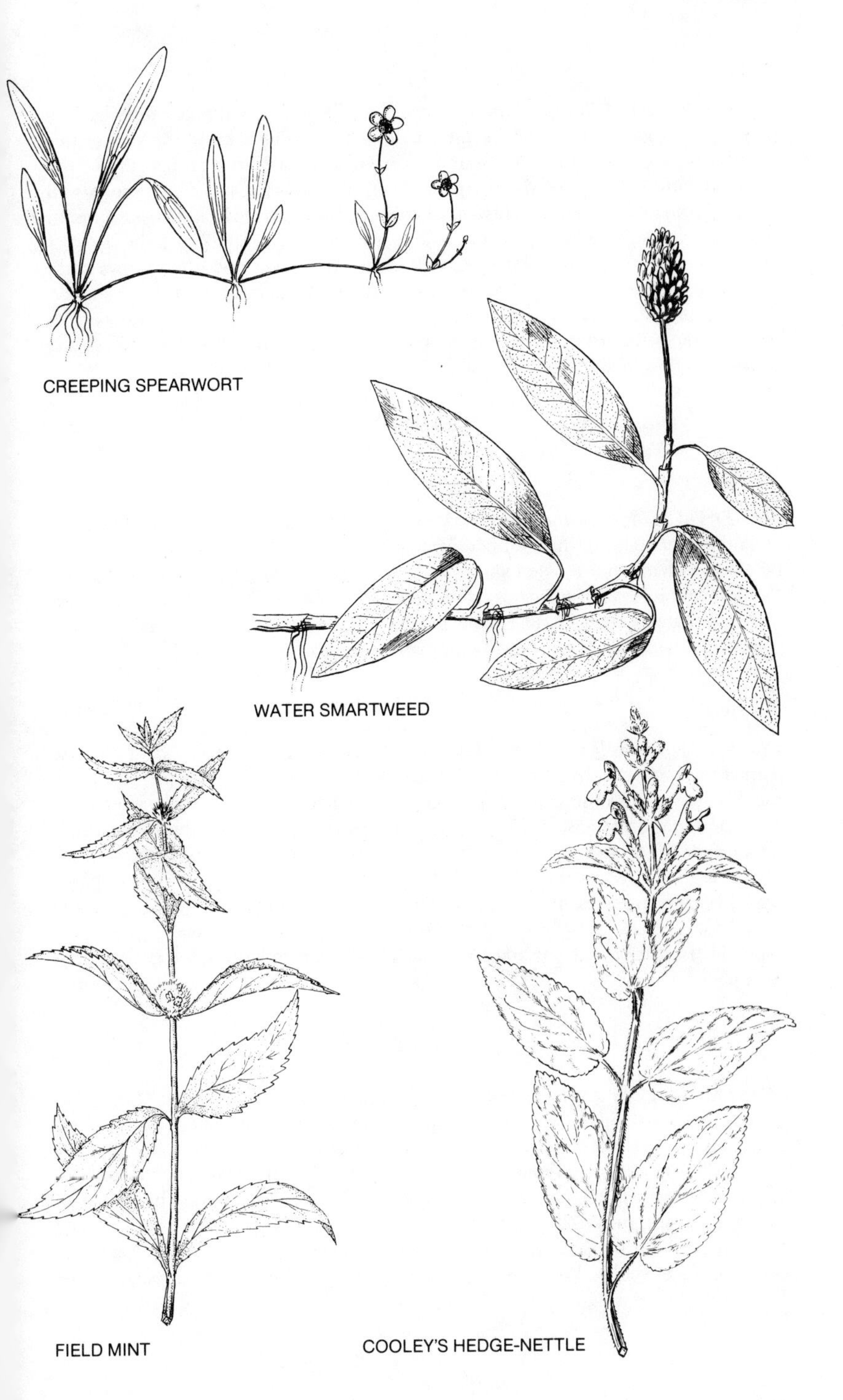
CREEPING SPEARWORT
WATER SMARTWEED
FIELD MINT
COOLEY'S HEDGE-NETTLE

Blue

SKULLCAP SPEEDWELL *(Veronica scutellata)*. When walking near the edge of a pond or lake, keep an eye out for the small, pinkish to light bluish flowers of this wildflower. When located, closer examination will reveal a distinct pair of anthers and a stamen protruding from the charming open-face flower. One of eight speedwell species found in the San Juans, this common one is distinguished by its flowers, present from June through August, and also its long, narrow leaves. Skullcap is the name of another wildflower, which this speedwell closely resembles and often grows with; speedwell is Old English for good-bye, though the association with the plant is unclear. To 10 inches tall. American speedwell *(V. americana)* is at least as common and occurs in much the same habitat, although it seems more numerous in shadier places and ditches. Its flowers are pale blue, and it has ovate, toothed leaves.

SHRUBS

Pink

HARDHACK *(Spiraea douglasii)*. In open, marshy places, this plant's pinkish plumes are a pleasant sight during summer. The common name is appropriate—just try to break one of the sinewy stems and you'll soon know why. Hardhack (also called Douglas spiraea) often forms dense thickets to 5 feet tall. The Latin species name commemorates David Douglas, the indefatigable Scottish botanist, for whom a considerable number of our Northwest plants were named.

Brown

SWAMP GOOSEBERRY *(Ribes lacustre)*. Shady lake edges, muddy places, or the damp forest floor are the preferred habitat of Swamp gooseberry, a common species in the San Juans. Blooming in spring, the saucer-shaped flowers are an agreeable cinnamon color, arranged in drooping racemes. More apparent are the stems, which are uniformly covered with small spines and have three–seven larger ones at each node; the growth form varies from vining to shrubby. Note the maple-leaf shape of the small, dark green leaves. In quantity, the tart-tasting black berries are used in jams and jellies. Coast black gooseberry *(R. divaricatum)* is a sturdier shrub of drier sites, especially on forest margins and rocky outcrops. A single large spine is present at each node.

TREES

PACIFIC WILLOW *(Salix lasiandra)*. Although distinguishing among the various willows can prove difficult, a few are easily recognized, as is this one. Because of the long, glossy leaves, and coarse glands near the leaf bases, Pacific willow looks unlike any other willow in our area. Taller than most of the Northwest willows, it may reach 80 feet or more in height. Dark, furrowed bark covers the crooked trunks, which favor an ascending to upright shrubby form; the wood is of little value because it is very soft and burns poorly. During spring, the bright yellow male catkins are obvious; later they become white-woolly. Very common in or near ponds and lakes throughout the San Juans.

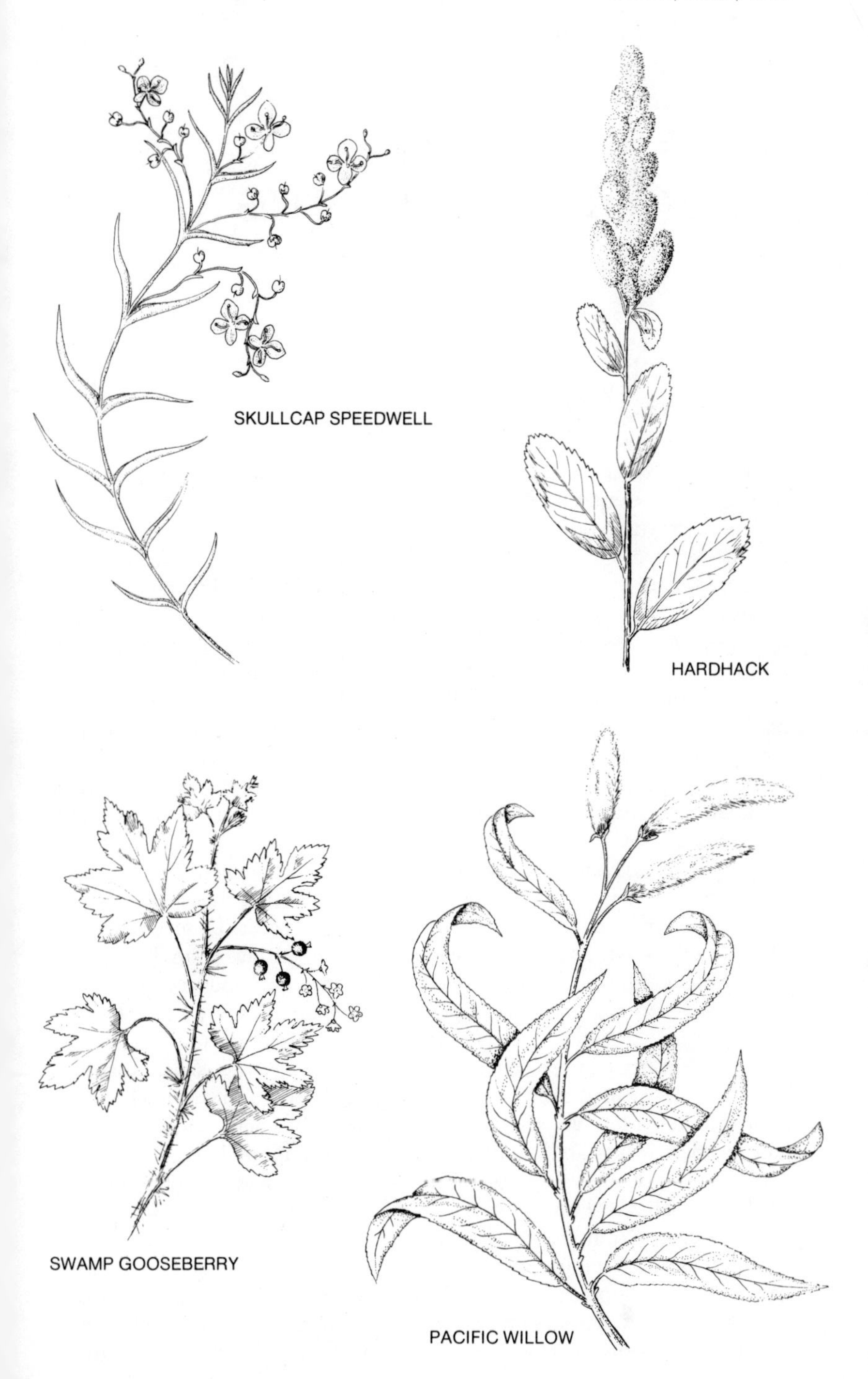
SKULLCAP SPEEDWELL
HARDHACK
SWAMP GOOSEBERRY
PACIFIC WILLOW

6
DISTURBED SITES

Although generally less inspiring than the native wildflowers of the nearby meadows, the flora of weedy or disturbed sites constitutes a large and noticeable percentage of the total number of vascular plant species in the San Juans—approximately 30 percent—and, as such, must be addressed. Not all of these species are undesirable; among them are some of our best-known herbs and most attractive flowering plants. Their numbers can only be expected to increase with more human settlement. As might be expected, San Juan County's two largest settlements, Eastsound and Friday Harbor, play host to a great number of disturbed site species. Whether in town or outside, vacant lots, residential gardens, roadsides, and woodland edge provide a ready home for these pioneers, many of them weeds of Eurasian origin. Although the flora of every habitat is subject to some disruption, the true disturbed site species is one that occurs consistently in areas that receive frequent, or an inordinate amount of, alteration. However, species associated with pristine areas will sometimes appear at the roadside ditch also and, of course, the opposite is common, i.e., disturbed site adventives will often dominate comparatively undisturbed areas.

Subjected to poor, hard-packed soils, intense summer sunlight, and other hardships—such as trampling and uprooting—disturbed site flora is characteristically very hardy and aggressive. Colonization occurs by several means. Many of our weedy species are members of the Sunflower family, a group known for its highly specialized seeds, which readily disperse in the wind. Many species employ a sort of creeping or sprawling method, e.g., some of the cultivated garden plants; certainly this habit is not exclusive to them, for it is very common among fresh water and beach dwellers, as well as plants of other habitats. The "garden escapes," those cultivated species that sprawl beyond their garden borders, are often attractive, like the bluebells, and seem weedy only in their habits and habitat. Cultivation of imported species has also inadvertently introduced disturbed site species, in that seeds are often present on the rootball of exotics. Of course, there are several other vectors, such as birds, fill dirt, or camping equipment.

Because of their tenacity, certain pioneers may encroach on pristine habitats; the presence of the grass Yorkshire fog *(Holcus lanatus)* on rocky outcrop meadows provides an example. The dominance of the invaders over natives in such situations can often be attributed to a more oppressive growth habit combined with superior means of seed production and dispersal. Some species show little inclination toward extra foliage or height but show a considerable propensity for rapid multiplication, effectively forcing out competitors. Others develop thorns or poisonous seeds or foliage to avoid being eaten by herbivores. Adaptability is the key for disturbed site species. They can survive in habitats that are too harsh for native flora.

Himalayan blackberry *(Rubus discolor)* seems somewhat less common in the San Juans than elsewhere in western Washington, perhaps due to the dryness or isolation of the area. Its place has been filled by other weedy species, particularly grasses and composites. Various horsetails (particularly *Equisetum arvense)* and thistles (particularly *Cirsium arvense* and *vulgare)* are common. Queen Anne's lace, also called Wild carrot *(Daucus carota),* Buckhorn plantain *(Plantago lanceolata),* and Oxeye-daisy *(Chrysanthemum leucanthemum)* are among the most abundant species found at disturbed sites. Hairy cat's ear *(Hypochaeris radicata)* and Smooth hawksbeard *(Crepis capillaris)* are equally ubiquitous and may be confused with Common dandelion *(Taraxacum officinale).* The differences between these three yellow composites are clarified in the text. Sow-thistles (especially *Sonchus asper* and *oleraceus)* are common companions of various weedy grasses. Yorkshire fog, Orchard grass *(Dactylis glomerata),* and Cheatgrass *(Bromus tectorum)* are abundant.

FERNS AND FERN-ALLIES

FIELD HORSETAIL *(Equisetum arvense)*. Highly variable in form and habitat. Sterile green stems, typically scraggy and weak, are present from April until September. They may be erect (to 2 feet) or sprawling, each stem with whorls of spreading branches, which can be short and symmetric to long and irregular. Fertile stems appear early in spring, withering quickly; they are brown, branchless, and feature prominent sheaths at each segment. A whitish-fringed cone tops these thicker stems; this bears the spores. Abundant; dense clusters are frequent in deciduous woods, along roadsides, vacant lots, and similar sites. Giant horsetail *(E. telmateia)* is much larger and more upright, with occasional plants reaching 5 feet. Common in poorly drained or waste places. Breaking open the stem reveals a cylindrical, water-filled chamber. Dutch horsetail *(E. hyemale)* is the evergreen representative; it lacks branches. Locally abundant, forming dense colonies in moist places.

GRASSES AND GRASSLIKE PLANTS

COMMON RUSH *(Juncus effusus)*. Although everyone has probably seen hundreds of them, few probably realized that what they were looking at was a rush. This rush has round, very narrow stems that are pointed at the tips. Moist waste places are the habitat of this species, overwhelmingly the most common of a number of rushes in our area. The name rush originated in England, where rushes were used to make mats and ropes, for floor covering, and for other domestic uses, such as candles, called rush lights.

YORKSHIRE FOG *(Holcus lanatus)*. An abundant grass about disturbed sites, rocky meadows, or almost anywhere. Although unobtrusive in appearance, this is one of our most aggressive grasses, delighting in crowding out native flora or fighting among the roadside ruffians for "toughest kid on the block" honors. The spikelets are soft to the touch, and the awns are hooked; the most important field mark is the pubescence along the nodes of the culms. In many ways an even more intrusive grass is Cheatgrass *(Bromus tectorum)*, which is actually more common in many places than Yorkshire fog. The spikelets of Cheatgrass have long, straight awns and glisten in sunlight. The seeds readily stick to one's socks. Particularly abundant on open meadows of the west side of San Juan Island. The spikelets, if placed on the tongue, can easily work themselves down one's throat, by virtue of tiny spines, as one of your authors once discovered, much to his dismay.

FLOWERS

White

STINGING NETTLE *(Urtica dioica* var. *lyallii)*. All know this abundant species because sooner or later the stinging growth catches everyone unaware, leaving a burning sensation not soon forgotten. Surprisingly, the young leaves, when boiled, make a good tea and are an excellent potherb; in addition, a mild soup prepared by boiling nettle roots is said to be good. Northwest Indians used nettle fibre, which is quite strong, for fish nets and other uses. Nettles form dense colonies in disrupted woodlands, particularly of the alder or mixed type; also, they are numerous along roadsides and cut-over or grazed areas. Deep thickets to 5 feet high can render certain places impassable. Watch for the jagged-toothed leaves and, in the spring, drooping clusters of greenish white flowers borne in the axils of the leaves. A peculiar pungent odor arises from nettle thickets.

FIELD HORSETAIL

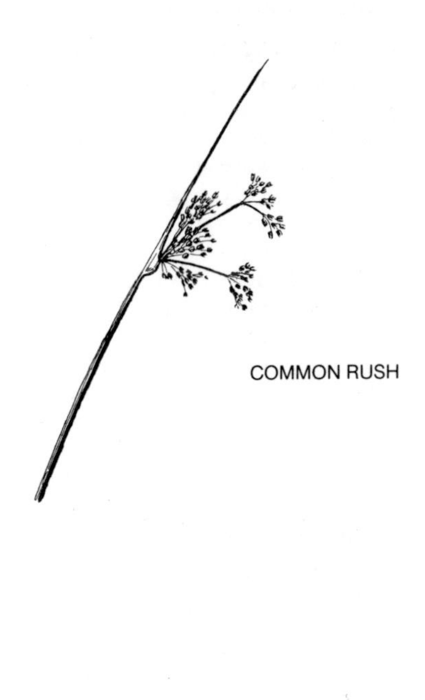

COMMON RUSH

YORKSHIRE FOG

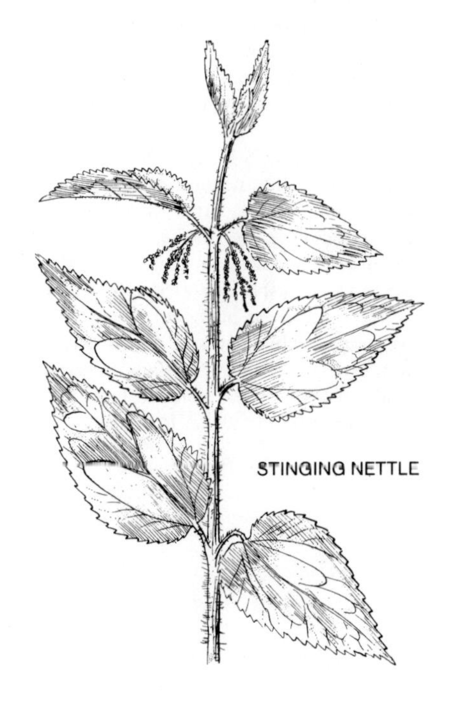

STINGING NETTLE

White (continued)

SMALL-FLOWERED CATCHFLY *(Silene gallica)*. A rugged weed of hardpacked waste soils, sunbaked meadows, and rocky outcrops. Long, stiff hairs cover the entire plant, although they are most noticeable on the tubular, inflated calyx. Shorter hairs form a fine pubescence in longitudinal lines below the larger hairs. From May through August small white or pinkish flowers appear. Each of the five petals has two toothlike appendages at the base, a good field mark. Common throughout the San Juans. European.

WHITE SWEET-CLOVER *(Melilotus alba)*. With a reputation for making good hay, for adding nitrogen to the soil, and for having a pleasant scent, this legume is certainly far more desirable than most of its roadside neighbors. It also supplies nectar for bees and seldom makes a nuisance of itself by overrunning an area. Easily recognized by the white-flowered spikes that sometimes reach 7 feet in height. Blooming from July to October, the flowers have a scent so sweet that it carries for some distance and, if taken up close or in excess, actually becomes rather sickening. When not in flower, it may be confused with alfalfa; however, White sweet-clover's leaves are toothed all along the edge and, when crushed, give off a pleasant scent. Common in the San Juans, but largely confined to roadsides. Yellow sweet-clover *(M. officinalis)* has yellow-flowered spikes, as the name suggests, but in other respects is very similar. Less common; it is restricted to sandy areas and roadsides, mostly on Lopez and San Juan Islands.

WHITE CLOVER *(Trifolium repens)*. Have you ever looked for a four-leaf clover, hoping to gain good luck? If so, chances are you were looking at the foliage of this clover, by far the best-known and abundant one. Foúnd world-wide, it is as abundant in the San Juans as elsewhere, growing on lawns, pastures, cracks in the pavement, etc. Most of us have watched bees busily attending the white flowers. Blooms from April until September, with flowers often fading to pinkish, or occasionally orangish, with age. The Latin species name *repens* means creeping; White clover uses runners to spread through lawns and other places. European.

QUEEN ANNE'S LACE *(Daucus carota)*. This plant is known as Wild carrot and believed to be the ancestor of our cultivated form. Although much tougher when eaten raw, Wild carrot is often sweet tasting. However, be sure not to mistake it for Poison-hemlock *(Conium maculatum)*, a much larger species with a purple-spotted or -lined stem and fernlike leaves. Poison-hemlock is a fairly common roadside weed in the San Juans, especially on San Juan and Lopez Islands. However, Queen Anne's lace is perhaps the most abundant weed in our area, and it is easily recognized by the flat-topped white umbels, which become cupshaped, like a bird's nest, with age. Rattlesnake weed *(D. pusillus)* is very similar but smaller, and it has smooth, short bracts rather than the elongate, hairy-spiny ones of *D. carota*. Prefers dry, rocky sites; fairly common in the San Juans.

SMALL-FLOWERED CATCHFLY
WHITE SWEET-CLOVER
WHITE CLOVER
QUEEN ANNE'S LACE

White (continued)

PEARLY EVERLASTING *(Anaphalis margaritacea).* The papery, parchmentlike flower heads, present from July to September, make identification easy. A whitish sheen radiates from the clusters, which usually have a golden center. The foliage is also distinctive: the long, alternate leaves are dark green above and light below, both sides white-woolly, as is the stem. Common and widespread along roadsides and meadows. The common name accurately describes a popular feature of the plant, that of retaining its beauty, without wilting, long after having been picked. The name also seems appropriate in light of the fact that it lingers along roadsides long after other wildflowers have withered.

OXEYE-DAISY *(Chrysanthemum leucanthemum).* Widely cultivated and readily escaping and spreading, Oxeye-daisy has now become a standard element of our roadside vegetation. From May to August, the large flowers are impossible to miss, each one with bright white petals (rays) and a yellow head. Another member of the Composite family with claims to medicinal value. Forms dense colonies in moist, disturbed places; cattle seem to avoid it. One of four chrysanthemums in the San Juans.

Yellow

FIELD MUSTARD *(Brassica campestris).* Dirt piles, farm fields, roadsides, and gardens will often show patches of this mustard. The bright yellow flowers may bloom at almost any time of the year but are usually present from May until September. The foliage is thick and fleshy and, if collected when tender, makes a good salad green or potherb. The field mark that distinguishes this from our other mustards is the smooth, blunt-tipped leaves clasping the stem. In farm country, dense growth often paints plowed lands yellow, as in Skagit County. Common in our area, particularly on San Juan Island. The European ancestor of cultivated turnip.

HEDGE MUSTARD *(Sisymbrium officinale).* A curious weed because of its growth form, with the seed pods closely hugging the branches. The branches themselves seem rather peculiar due to their spreading growth habit, which strikes one as ungainly and haphazard. During summer, tiny, pale yellow flowers are borne at the branch tips. Common around old farm buildings, moist grassy roadsides, vacant lots, and gardens. Tumble mustard *(S. altissimum)* has longer seed pods in an open arrangement not hugging the branches. The foliage is smooth, unlike that of Hedge mustard. Prefers drier roadsides and meadows. The dried stem framework gets blown around on windy summer days, a sight reminiscent of tumbleweeds, hence the name.

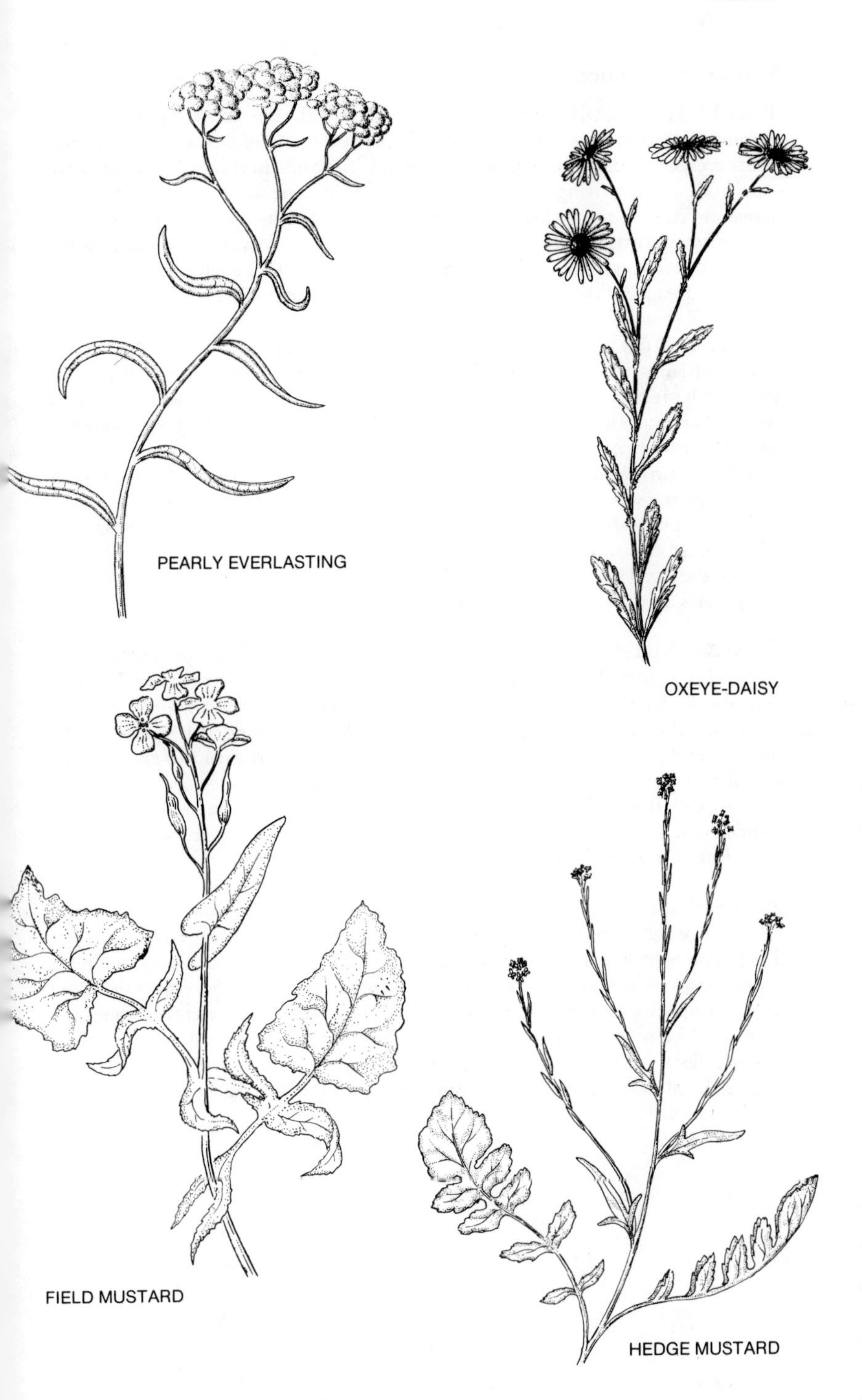
PEARLY EVERLASTING
OXEYE-DAISY
FIELD MUSTARD
HEDGE MUSTARD

Yellow (continued)

BIRDSFOOT-TREFOIL *(Lotus corniculatus).* Little dark green mats of leaves and low clusters of radiant yellow flowers are the simple field marks of this roadside resident, which is especially common on San Juan Island. The brilliant flowers are easily seen, even from a fast-moving car and, in fact, one may drive over them, as the mats frequently creep onto the pavement; blooms from May to September. Occasional reddish forms occur. The common name is a misnomer—the smooth leaves are usually in groups of five, not three, as the name trefoil suggests. Originally introduced from Europe as forage for cattle; common and apparently increasing.

ST. JOHN'S WORT *(Hypericum perforatum).* Truly an aggressive weed on Lopez and San Juan Islands, where it dominates several roadsides. Poisonous to stock, the plants are difficult to eradicate because of the rapidly spreading rhizomes. As with so many of our weeds, it came from Europe to North America where, without the particular beetle species that kept it in check, it spread like brush fire over most of the continent. Surprisingly, in medieval Europe it was thought to be effective in warding off illness or bad fortune and was traditionally burned in bonfires on June 24—the feast day of Saint John the Baptist—hence the name. Plants routinely reach 2–3 feet in height. During mid-summer the showy clusters of large, bright yellow flowers become obvious. Because the petals lie flat atop each other, the 50–plus stamens in each flower are conspicuous. Also known as Klamath weed.

YELLOW BARTSIA *(Parentucellia viscosa).* The famous Swedish botanist Linnaeus named this plant for his friend Johann Bartsch; the genus name comes from Tomasso Parentucelli, former curator of the Rome Botanical Garden. Similar to the Monkey flower and closely related, Yellow bartsia is often found growing with it. However, it may be distinguished by the faded yellow flowers, which feature a trilobed lower lip and sack-shaped upper one. From April to September, the flowers are borne in terminal racemes, each flower bracted by clasping, toothed, fleshy leaves. The foliage tends to be glandular-pubescent. Stems to 2 feet. Partially parasitic on nearby grasses and weeds. Abundant in wet open places, particularly on the larger islands.

FLANNEL MULLEIN *(Verbascum thapsus).* The mammoth spikes, each several inches thick and up to 6 feet high, are perfectly straight and linger, dark brown and dried, long after the yellow flowers are gone. When spikes are absent, the huge soft-hairy leaves are obvious, even from a distance. When dried, they sometimes are used as a smoking mixture or as a tea. Each of the large flowers starts blooming at a slightly different time, between June and September. Not as aggressive as most roadside flora, tending to appear only in places where grasses and other weeds are much reduced. Scattered but fairly common along roadsides, rocky areas, and meadows in the islands.

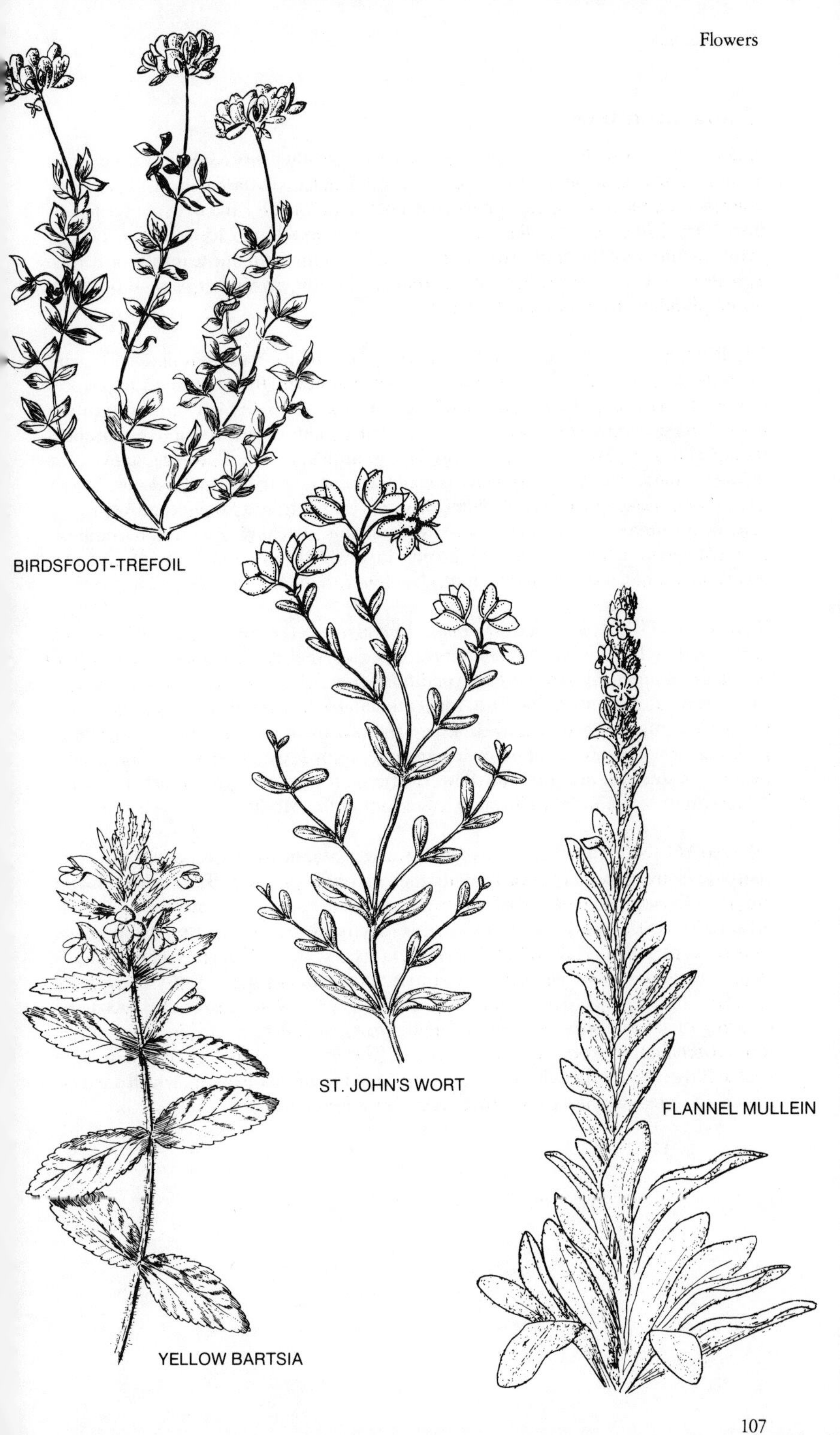
BIRDSFOOT-TREFOIL
ST. JOHN'S WORT
FLANNEL MULLEIN
YELLOW BARTSIA

Yellow (continued)

SMOOTH HAWKSBEARD *(Crepis capillaris)*. Virtually every roadside, vacant lot, or meadow in season displays Smooth hawksbeard, abundant in the San Juans. The yellow flowers resemble those of Common dandelion but are smaller and in dense clusters of five–thirty. Hawksbeard's stems are also taller, often reaching 2 feet and, like those of many members of the Sunflower family, exude a milky juice if one tears off the leaves. The deeply divided, pinnatifid leaves are unlike those of any similar plant of the same habitat. Blooms from June well into fall.

HAIRY CAT'S EAR *(Hypochaeris radicata)*. Here is the dandelion look-alike, and many persons are fooled by it. The flowers seem identical to the dandelion's at a glance; however, the leaves are quite different. They are thick, very hairy, dark green, and often matted. People who have lawns know the matted clumps well, for once established, they quickly join plantains in crowding out grass. In addition, the flowering stems seem to have a unique ability to pass under lawn mowers without being cut. If Queen Anne's lace is our most abundant weed, then Hairy cat's ear has to be a close second, occurring at every disturbed site and meadow in the San Juans. Actually, it is far more common than dandelion because it tolerates greater dryness. Blooms from May into fall. An excellent wild salad source, being much less bitter than Common dandelion.

PRICKLY LETTUCE *(Lactuca serriola)*. Few plant species can survive conditions as harsh as those Prickly lettuce endures. Hard-packed, sunbaked soils, seemingly devoid of nutrients support this weed. Such conditions exist in Friday Harbor, where cracks in the pavement, vacant lots, and road edges are suitable for growth. Some people prefer to call it Compass plant, a name based on the way the leaves jut out from the stem, pointing one direction or another. These large, veiny leaves are the most noticeable feature because the insignificant flowers often fail to open. Tearing off the leaves brings an instant ooze of milky juice. Abundant in the San Juans.

OLD MAN-IN-THE-SPRING *(Senecio vulgaris)*. Seldom over 6 inches tall, this plant is apt to be overlooked by most, even though it is quite common. The key field marks are the yellow flower heads that lack petals and have black-tipped bracteoles, and the deeply divided leaves. Found in rocky meadows, moist or waste places, gardens, and forest clearings. Wood groundsel *(S. sylvaticus)* is very similar but taller, and the heads have a few tiny yellow petals. Less common in similar locations. Tansy ragwort *(S. jacobaea)* has become quite common in plowed fields, waste places, and woodland clearings. It is poisonous, and efforts to eradicate it, particularly in farming areas where it has been fatal to stock, are on the increase. The key field marks are the many large yellow flowers atop the tall (to 6 feet) stems, and the alternate leaves, which are dark green and irregularly lobed. All three species are European.

SMOOTH HAWKSBEARD

HAIRY CAT'S EAR

PRICKLY LETTUCE

OLD MAN-IN-THE-SPRING

Yellow (continued)

PRICKLY SOW-THISTLE *(Sonchus asper).* Stiff, shiny, prickly leaves are characteristic of this European invader, very common in waste places throughout the islands. During July–October, dark yellow flowers, to 2 inches long, are terminally borne. Distinguished most easily from other sow-thistles by the darker, stiffer leaves. Common sow-thistle *(S. oleraceus)* is about equally common in the same habitats. It has light green, softer, less prickly leaves. Both species have highly variable leaf shapes. The foliage of both species is edible.

COMMON DANDELION *(Taraxacum officinale).* Everyone knows dandelion. This, perhaps the best known of our flowers, has proven to be much more than just a garden weed. Excellent wine is derived from the flower heads, and the young shoots make good salad greens. Dandelion roots are reputed to have medicinal value. The name comes from the French *dent de lion*, making reference to the ragged outline of the leaf margins. A handsome flower whether in shaggy golden bloom or gone to seed in pale globes. Compare this species with Hairy cat's ear (p. 108).

Orange

POOR MAN'S WEATHERGLASS *(Anagallis arvensis).* A distinctive species restricted chiefly to San Juan Island in our area, where it is common along several roadsides. An interesting plant with an intriguing common name, derived as a result of observation made in England that the flowers close when the clouds roll in. During full sun, the flowers open wide, pointing skyward, to reveal the appealing orange petals and purple centers. The weak, prostrate stems seldom exceed 12 inches in length. This Eurasian immigrant blooms during summer. Also called Scarlet pimpernel.

Red

CURLY-LEAVED DOCK *(Rumex crispus).* A familiar sight lasting from summer well into winter is the dried, dark brown spikes of various docks, obvious along roadsides and fields. Several species are common in the San Juans, but this one seems most abundant, turning up along seashores, forest edges, and pastures. The key field mark of this dock is the large basal leaves, which are curly or ruffled along the edges. Tiny, reddish bell flowers are borne in the leaf axils in spring. By June or July they are replaced by equally small, dark brownish seeds, which are rounded, without barbs, and form dense clusters. The tender, young leaves may be boiled and eaten like spinach. Eurasian.

PRICKLY SOW-THISTLE

COMMON DANDELION

POOR MAN'S WEATHERGLASS

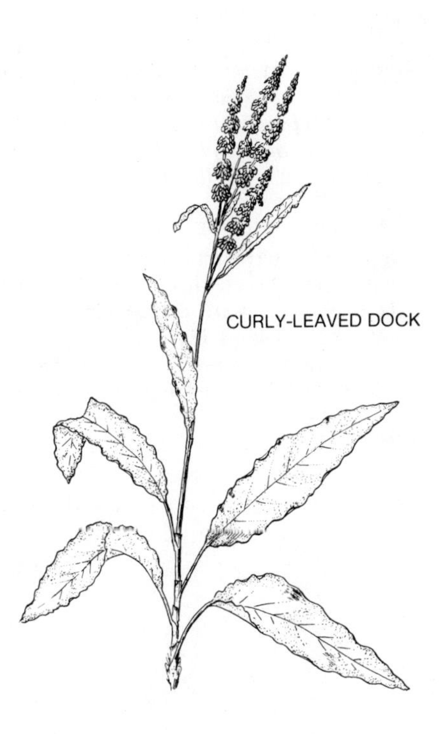

CURLY-LEAVED DOCK

Red (continued)

RED CLOVER *(Trifolium pratense)*. Although many clovers are known from the San Juans, this, and the White clover, are overwhelmingly the most abundant. Red clover has dark green leaves, each with a prominent white or dark crescent near the center; the upper leaves sometimes form a false involucre around the flowering heads. A multitude (50–200) of tiny pinkish or reddish flowers make up each robust head. Eurasian.

Pink

DOVEFOOT GERANIUM *(Geranium molle)*. Most persons are apt to notice the rounded, lobed leaves (1–2 inches) before they see the stems, seeds, or flowers, which are often inconspicuous among the heavy grass cover of the roadside or lawn. However, the pink or magenta flowers will sometimes be quite noticeable atop a 12-inch stem. The flowers appear erratically from May to August. After blooming, the linear seed pods, called carpels, appear. This species is distinguished from the other geraniums by its smooth carpels (although they have tiny spines along the segment lines). Very common, especially on San Juan Island. The elongate carpels have earned it and other members of the family another name: Crane's bill. European.

FIREWEED *(Epilobium angustifolium)*. In light of this wildflower's many attributes, it seems that it is anything but a weed. Widespread across North America and Eurasia, it has provided inspiration for writers, tea drinkers, natural foragers, and beekeepers. From June to September, the grand pinkish purple flowers bloom from the base upward. When blooming is finished, the silky white seed parachutes appear, a beautiful sight as they glint and glimmer in the autumn sunlight. Fireweed tea, made from the dried leaves, is especially popular in the USSR, where it is called kaporie tea. The use of young shoots as a potherb has been more popular in this country. Fireweed honey is also highly prized. This species is invariably among the first pioneers at a burned-over area, often growing in profusion. In the northwest, we also associate it with cut-over areas, roadsides, and forest margins as well. Common throughout the San Juans.

WATSON'S WILLOW-HERB *(Epilobium watsonii* var. *watsonii)*. Though no match in beauty for the closely related Fireweed, this species is even more common, occurring in moist places everywhere in our area, in full sunlight or shade. Fleshy best describes the foliage; leaves are toothed and 1–4 inches long. The free-branching stems may reach 2 feet. During July and August, watch for the tiny pink flowers borne at the branch tips. Each flower has four notched petals. Stems and leaves often have a deep reddish color. Does well in gardens, roadside ditches, and woodland margins.

RED CLOVER

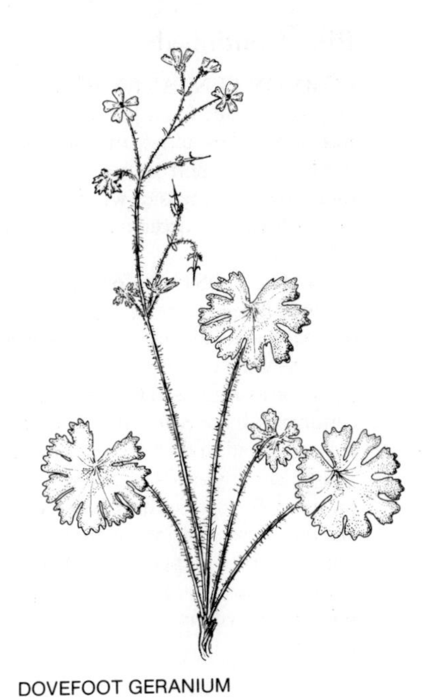

DOVEFOOT GERANIUM

FIREWEED

WATSON'S WILLOW-HERB

Pink (continued)

COMMON CENTAURY *(Centaurium erythraea).* Flat, showy pink "stars" develop atop each plant from July to September. Often forming substantial colonies along roadsides and meadows, this species is by far most abundant on San Juan Island, though it is locally common on other major islands as well. Erect stems reach 2 feet. One of the more attractive roadside weeds. Apparently it is near its northern limit in the San Juans and Gulf Islands, occurring south to California.

CANADA THISTLE *(Cirsium arvense* var. *horridum).* The Canada or Field thistle is horrible indeed, among the most hated and difficult to eradicate of the weeds. An outlaw, this thistle had U.S. legislation appear against it as early as 1795, a short time after the weed was introduced, apparently accidently, by Canadian settlers. Originally a native of southeastern Europe and adjacent Asia, it is now an intense nuisance on every continent except Antarctica. Hacking off foliage and even cutting the roots often only aids its spread, and wherever heavy grazing, clearing, or other disruption occurs, there is little doubt Canada thistle will follow. At Cattle Point, San Juan Island, and other places, native species have been largely choked out by it. If one hoes it out, root and all, and then burns everything, maintaining a vigil over the area for one–two years, one can hope it will not reappear. Bull thistle *(C. vulgare)* is another European adventive that is scarcely more desirable but is slightly less common. It has more deeply cleft leaves, seldom forms dense groupings as *C. arvense* does, and has reddish purple, not pinkish, blooms.

Purple

ALFALFA *(Medicago sativa).* Also known as Lucerne. Often cultivated, it is widely escaped in the San Juans, growing commonly along roadsides, meadows, and in weedy places generally. Hard to mistake for anything else: the small purplish flowers bloom over much of the growing season, and each of the three leaflets (1–3 inches long) are toothed at the tip. The seed pods are curled up in a snaillike fashion. Stems are low and spreading, to 2 feet or so. This is another European species planted as forage; it is also important as a source of nectar for honeybees.

PURPLE CUDWEED *(Gnaphalium purpureum).* Several cudweeds are found in the San Juans; this one has the distinction of being somewhat less weedy than the others in appearance and habitat. Although it occurs in waste areas, Purple cudweed may also be found on forest margins and relatively undisturbed meadows, growing among native herbs. The gray foliage and stem, covered with woolly hairs, are as prominent as the dense inflorescence, which is purple during its brief summer blooming period and brownish at other times. Note that the alternate leaves are largest at the base and that plants will often grow in clusters of three. Usually 6–12 inches tall.

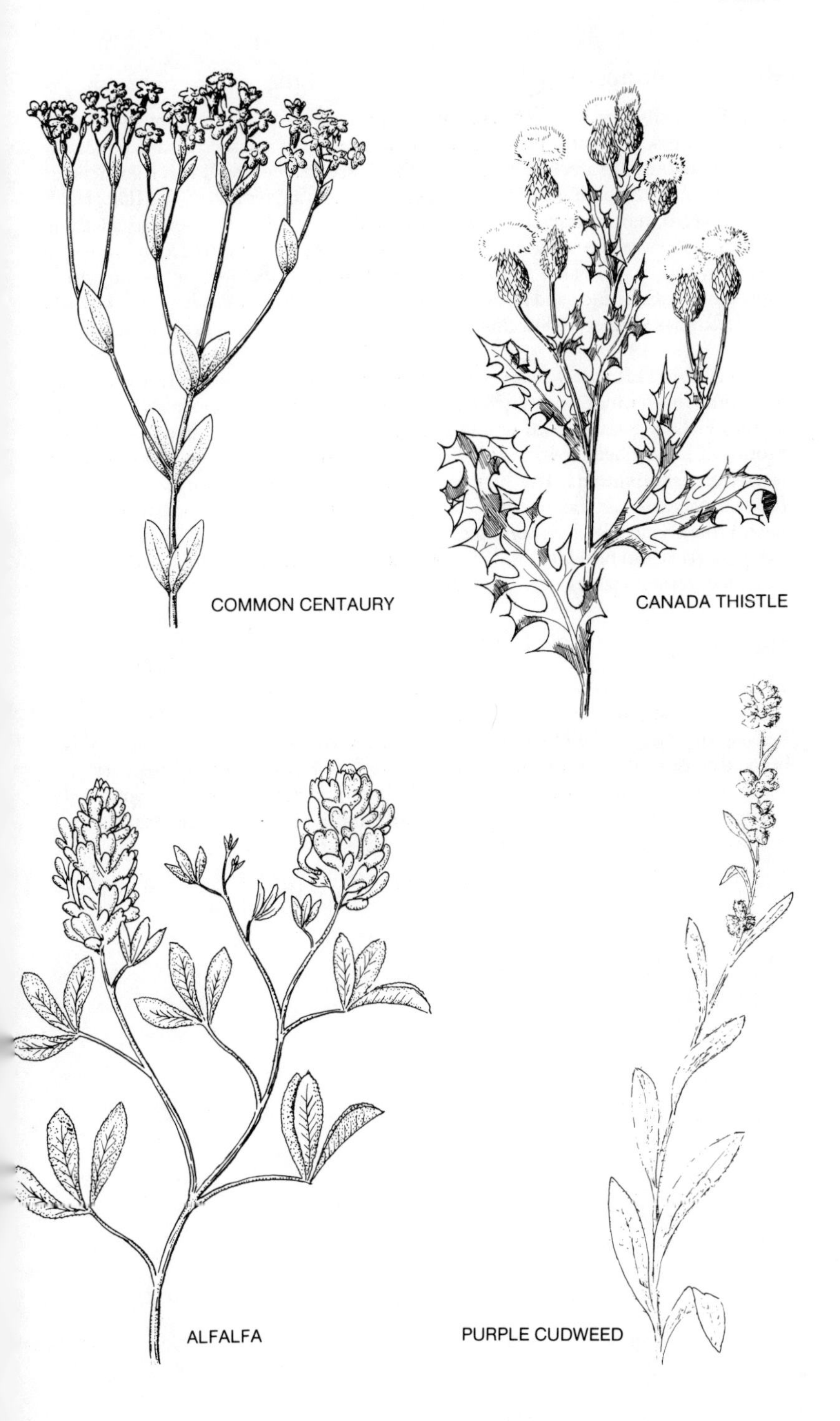
COMMON CENTAURY
CANADA THISTLE
ALFALFA
PURPLE CUDWEED

Purple (continued)

SALSIFY *(Tragopogon porrifolius)*. A conspicuous species because of its striking purple flowers. Atop the thick stems, the flowers have petals that are noticeably shorter than the light green bracts that surround them. Sometimes a rich golden center is present. After blooming during June–July, the flowers are replaced by big fluffy heads of aerial seeds, reminiscent of dandelion but several times larger. The wavy, clasping leaves are grasslike and edible when young. The roots too can be eaten; their likeness in taste to oysters has given rise to another name, Oyster plant. Salsify is apparently an old French name. Cultivated and fairly regularly escaping on our roadsides. Yellow salsify *(T. dubius)* has smaller, yellow flowers and is less common; found in the same habitat.

CHILEAN ASTER *(Aster chilensis)*. Along many roadsides in the San Juans one is apt to run into this harmoniously attractive species, as it frequents wet to fairly dry ditches as well as lake edges, disturbed sites in general, and, less commonly, maritime habitats. From July to October showy purple flowers are evident atop 2–5-foot stems, which often form dense clusters. The 20 or more petals (rays) often curl backward, as do the fleshy involucre bracts below them. This is one of the most abundant asters is Puget Sound country and is often confused with other similar species. Douglas' aster *(A. subspicatus)* is also present in the San Juans, although it is less common and generally restricted to moist places near salt water.

Blue

BLUE AND YELLOW FORGET-ME-NOT *(Myosotis discolor)*. Both light blue (older) and yellow (younger) flowers are present during spring, but they are inconspicuous due to their minute size (⅛–¼ inch wide). Scattered along the stem, they become densest near the top, where the stem curls to form a tight inflorescence. The short stems bear alternate leaves. A weedy Eurasian annual found throughout the archipelago on rocky or hard-packed soils and disturbed places. One of several *Myosotis* species in the area, but the only one with flowers of two colors.

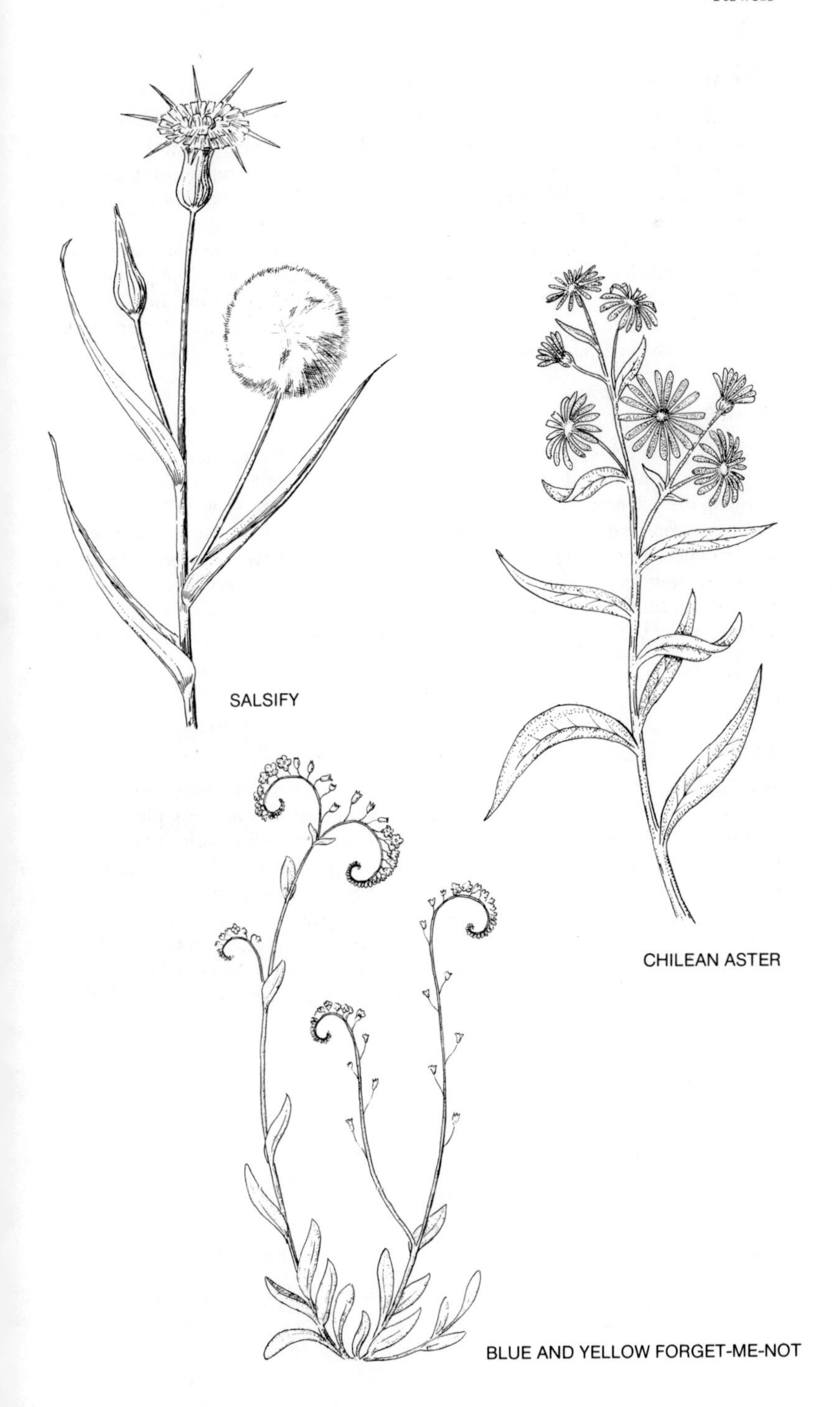
SALSIFY
CHILEAN ASTER
BLUE AND YELLOW FORGET-ME-NOT

Blue (continued)

BLUE SAILORS *(Cichorium intybus).* What fine sky-blue flowers this species has! Blooming from July until October, the silver-dollar-sized blooms will close up with the onset of bad weather, even at mid-day. Some Europeans have apparently valued Blue sailors' root as a coffee adulterant more than Americans have; in fact, the response to it in the U.S. has often been one of disgust. Attempts to establish Blue sailors here as a regular farm crop during the late nineteenth and early twentieth centuries proved a failure. Nonetheless, farmers value it because it makes good hay; some persons have found that the young leaves make a good potherb. Also called Chicory, Blue daisy, and Succory. The angled stems may reach 4 feet, and in dried form they linger through winter. Mediterranean; common in the San Juans.

Brown

BUCKHORN PLANTAIN *(Plantago lanceolata).* A rather nondescript weed, which all persons trying to maintain a lawn or garden will encounter. As with many weeds, cropping the dull greenish leaves or 12-inch stems will hardly slow it down; one must remove it roots and all. A European species now found on every continent and abundant here. Common plantain *(P. major)* is also abundant but has longer heads and stemmed, ovate leaves. Var. *major* is found in waste places and wet sites throughout; it has smooth, thin leaves. Var. *pachyphylla* is restricted to saline habitats such as the rocky or sandy beaches on San Juan and Orcas Islands. It has hairy, succulent leaves.

SHRUBS

Yellow

SCOTCH BROOM *(Cytisus scoparius).* This immigrant from the British Isles has been as successful in colonization as its human counterparts who first planted it on Vancouver Island; now it is widespread over coastal areas from British Columbia to California. The bright yellow flower clusters, appearing May–July, contrast handsomely against the dark green stems and foliage. Occasionally one finds striking reddish tinged flowers (var. *andreanus).* The seed pods begin splitting open during August. Their crackling, combined with the buzzing of grasshoppers in flight, is a familiar sound on hot summer days in Puget Sound country. Nonetheless, one must remember that this shrub's rapid spread has come at the expense of many desirable native species. Common in the San Juans.

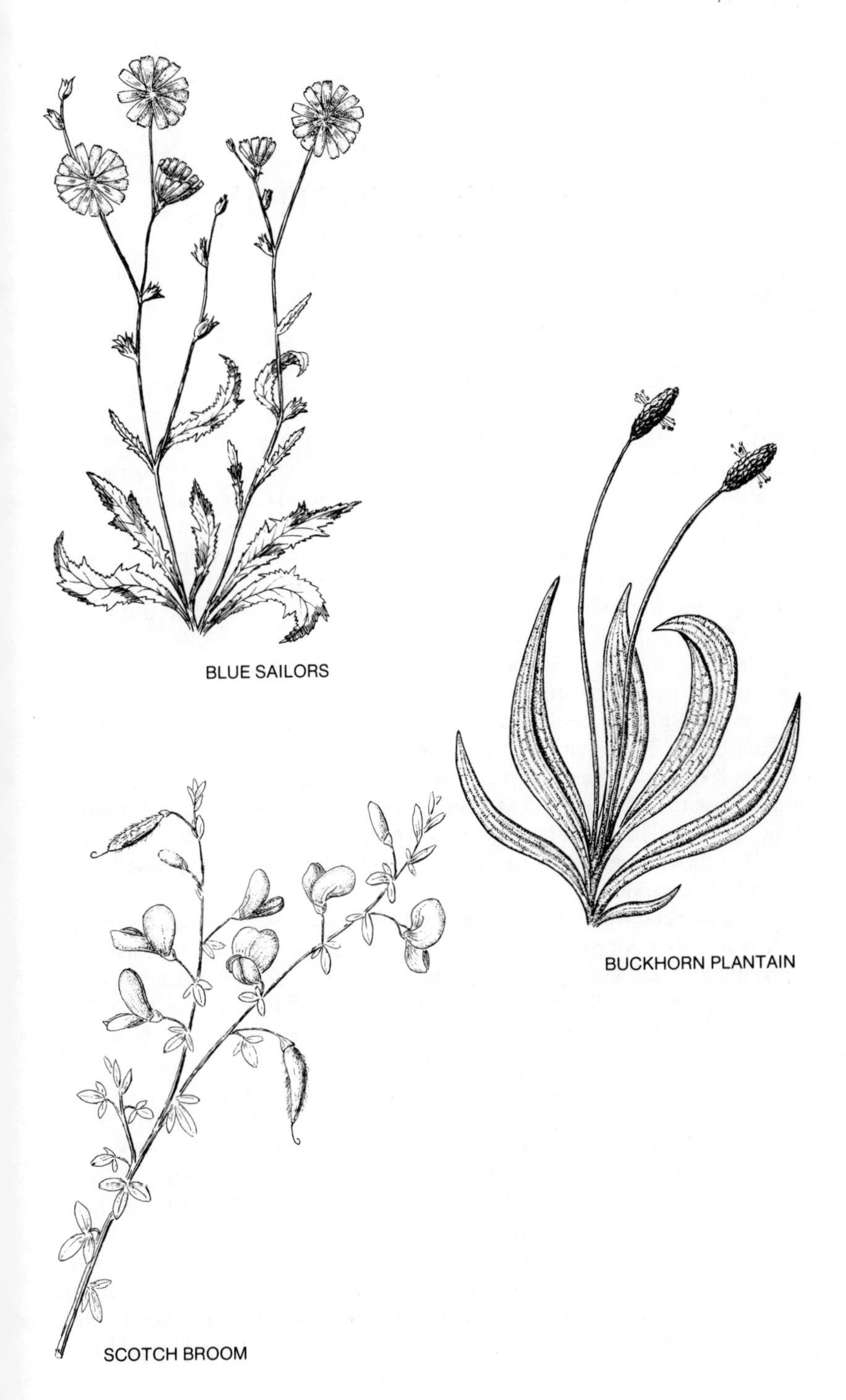
BLUE SAILORS
BUCKHORN PLANTAIN
SCOTCH BROOM

7
MT. CONSTITUTION

The flora here is unique in that it includes several species rare or absent elsewhere in the archipelago, and for this reason it is treated in a special section. Several species of a markedly northern character are noteworthy. The summit area, rising to 2409 feet in elevation, is high enough to support several mid-montane and subalpine species of the Cascades and Olympics, and were it not for the mountain's isolation and comparatively dry conditions, it would likely have more of these. An uncommon combination of dryness, higher elevation, and San Juan influence, joined with the mountain's isolated Puget Trough position and various geologic factors, produce a distinctive, interesting flora.

Mt. Constitution is the high point of an erosion-resistant band of sedimentary rocks, mostly siltstone and graywacke. The last glaciation left the mountain with a plateaulike area above 2200 feet, and with several perched water tables. The resulting poor drainage is evident today in several shallow, boggy lakes and sedge meadows. These swampy habitats, while ostensibly less appealing than Cascade and Mountain Lakes down lower on the west flank of the mountain, have few botanic counterparts in San Juan County. Therefore, flora of these boggy places and of the immediate summit is the primary focus of this description.

Colder and wetter than those at sea level locations, the summit's climate plays an important role in the floral composition. Approximately 45 inches of precipitation fall annually, a larger percentage of it falling as snow than at lower elevations. Although this is a lower total than at corresponding elevations in the Cascades and Olympics, it is more than double the annual average received on southern Lopez Island. The summit's July average temperature, approximately 55° F, and the January average, approximately 32° F, are about 5°–7° F lower than those at sea level. The spring bloom arrives as much as a month later and, similarly, killing frosts arrive earlier. Higher winds and humidity are also characteristic. Such montane conditions are approached at a few other San Juan locations, but nowhere is the effect so pronounced as at Mt. Constitution's summit.

Ferns are of particular interest on Mt. Constitution, with three noteworthy species present. Two of them are small, tufted types: Indian's dream *(Aspidotis densa)* and Rocky Mountain woodsia *(Woodsia scopulina),* both growing in rock crevices. Indian's dream grows in the Cypress–Fidalgo Island group on serpentine outcrops but is absent from San Juan County except for the Mt. Constitution population; Rocky Mountain woodsia is restricted in San Juan County to Frost Island and Mt. Constitution. Maidenhair spleenwort *(Asplenium trichomanes)* grows inconspicuously on some south-facing rock seeps at about 1700 feet, next to the main road up to the summit; it is very scarce on San Juan Island also. Oregon stonecrop *(Sedum oreganum)* is common in similar habitats. Spotted, Tufted, and Western saxifrages *(Saxifraga bronchialis, caespitosa,* and *occidentalis)* occur here as well. The dense Lodgepole pine woods south of the summit are host to Dwarf mistletoe *(Arceuthobium tsugense),* a peculiar parasite that normally attacks hemlock. Scattered along the trails around the summit is Varied-leaved collomia *(Collomia heterophylla),* a small species rather infrequently found elsewhere in the San Juans. Showy polemonium *(Polemonium pulcherrimum)* is more conspicuous along open rocky slopes. Often found near the summit is Rosy pussy-toes *(Antennaria microphylla),* unknown elsewhere in the San Juans. Dwarf mountain daisy *(Erigeron compositus)* grows chiefly near the summit on exposed rocky slopes. Puget butterweed *(Senecio macounii)* provides splashes of yellow at the summit and along the road up. Rayless mountain butterweed *(S. indecorus)* grows at Summit Lake.

FERNS AND FERN-ALLIES

MAIDENHAIR FERN *(Adiantum pedatum)*. Beaded with droplets or quivering from the nearby spray, this distinctive fern revels in its misty, moss-covered bank environment, usually next to a waterfall or stream. The tidy, handlike symmetry, the stems' shiny brownish or jet-black luster and water-resistant, lobed pinnules (leaflets) make this fern unmistakable. A familiar, beloved fern the traveller will often encounter crossing a mountain stream tumbling through the rocks. In the San Juans, the modest but appealing waterfalls above Cascade Lake on Orcas Island are a dependable location for finding it. Also occurs in rocky seeps and on moist streambanks on and near Mt. Constitution; rare on San Juan Island.

INDIAN'S DREAM *(Aspidotis densa)*. A tufted denizen of the rock crevices, occurring chiefly on Mt. Constitution in the San Juans. Blackish, wiry, and lustrous, the stems (to 12 inches) seem as if they've been polished. Note the narrow, pointed pinnules. Particularly abundant on the Cold Springs trail, where, in a few meadows, it is nearly the dominant species. Also found on Cypress, Fidalgo, Allan, and Burrows Islands east of the main archipelago, where it prefers dry prairies on serpentine substraits. Parsley fern *(Cryptogramma crispa)* occurs in the same habitat and is quite similar. However, the fronds are more densely clustered and both fertile and sterile fronds are often present; both types have blunt-tipped pinnules. More widespread (although scattered) than Indian's dream.

MAIDENHAIR SPLEENWORT *(Asplenium trichomanes)*. Discreetly growing in the wet, mossy rock crevices, this little fern will only be detected by the wary observer, despite its striking spring-green fronds. The leaflets are similar to those of Maidenhair fern, hence the name. The jet-black stems are seldom over 6–8 inches long. Occurs very sparsely on Mt. Constitution; also occurs on San Juan and Blakely Islands, although rarely.

ROCKY MOUNTAIN WOODSIA *(Woodsia scopulina)*. Rocky crevices between Little Summit and the Stone Tower Lookout are the best places to find this fern, another species occurring almost exclusively on Mt. Constitution. In fall, fronds fade to yellowish or brown, attractive as they grow from the cliff crevices. The leaflets are hairy and rather distant from each other. Occurs more commonly east of the Cascades.

MAIDENHAIR FERN

INDIAN'S DREAM

MAIDENHAIR SPLEENWORT

ROCKY MOUNTAIN WOODSIA

FLOWERS

White

LONG-STALKED STARWORT *(Stellaria longipes* var. *altocaulis)*. Among the rarest of San Juan species, this starwort has only been found at a single wet meadow area on Mt. Constitution, despite our searches to locate it elsewhere. The showy white flowers, 3/4–1 inch wide, are larger and far more conspicuous than those of other *Stellaria* species. Snaking through wet grasses, the stems (up to 2 feet) are low and curving; they support stiff, pointed leaves, an excellent field mark. The genus name is derived from the Latin *stella,* or star, a reference to the neatly star-shaped flowers of the genus.

BROAD-LEAVED SUNDEW *(Drosera rotundifolia)*. Who is this peculiar little fellow of the sphagnum mat, who has boldly taken up the practice of its southerly cousins, the Venus fly-traps, in devouring the luckless insects that land on it? Whose little basal tuft of spoonlike, glandular-hairy leaves is this that invariably arouses our interest as we visit the bog? Broad-leaved sundew is the carnivore with the basal tuft of leaves, luring its victims by virtue of sweet, sticky globules at the tip of each hair. When an insect lands on the leaf, it immediately becomes caught and the leaf curls up around it. From May to August, tiny snow-white blooms rise above on 4–12-inch stems. Found over much of North America and Eurasia. In the San Juans, most common on Orcas Island, but an extensive colony is at the Beaverton Marsh, San Juan Island, and the species is also known from Blakely Island. The small islets in Summit Lake are a good place to find it on Mt. Constitution.

SPOTTED SAXIFRAGE *(Saxifraga bronchialis* var. *austromontana)*. A bewitching beauty in miniature, each of its flowers has a harmonious symmetry. Stamens fit evenly between each of the shiny white petals, themselves adorned by tiny purple speckles. Blooming from June to August, many flowers are present in an open panicle 2–12 inches high. When not blooming, note the dense basal tuft of needlelike leaves, each one linear, pointed, and spiny along the edge. Found mostly near the summit in wet meadows and rock crevices. Tufted saxifrage *(S. caespitosa* var. *subgemmifera)* also occurs on Mt. Constitution, as well as several other Orcas and other island locations. It has tiny, mostly basal leaves that are trilobed into a rabbit's-foot shape, has unspotted white petals, and has hairy foliage, unlike that of Spotted saxifrage.

ROSY PUSSY-TOES *(Antennaria microphylla)*. Appropriately named, the round, rosy whitish heads do indeed resemble the toes of a pussycat. A low, runner-bearing species of stony ground near the summit, seldom over 10 inches high. Each flower has papery, rosy bracts around the white center. The evergreen leaves and stems are covered with soft, white hairs. Blooms in summer. A typical species of the Cascades and Olympics, disjunct on Mt. Constitution.

LONG-STALKED STARWORT

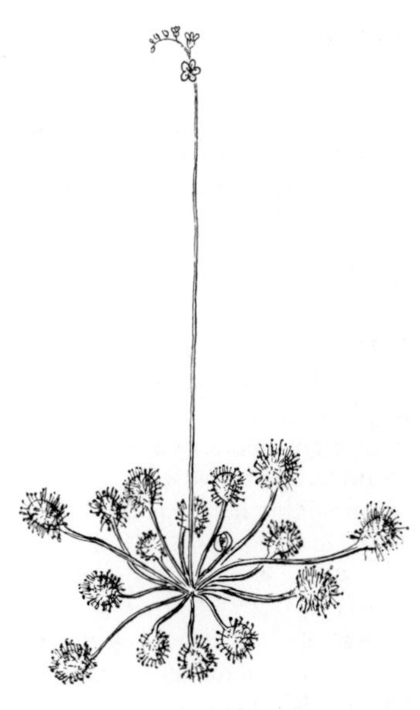

BROAD-LEAVED SUNDEW

SPOTTED SAXIFRAGE

ROSY PUSSY-TOES

Yellow

OREGON STONECROP *(Sedum oreganum)*. Hugging the rock crevices with other small herbs and ferns, the thick rosettes of this stonecrop are a deliciously burgundy-red color. Restricted in the San Juans to Mt. Constitution above approximately 1500 feet, this plant is primarily a montane species of the Cascades and Olympics in Washington, disjunct in the San Juans. Radiant golden yellow flowers rise above some, but not all, plants in summer, adding a fine decor to the steep slopes. Each flower has five narrow, pointed petals that are basally fused.

FRINGED PINESAP *(Hypopitys monotropa)*. This dark-woods dweller seems like a golden ghost, emerging eerily from the deep humus. On first impression the observer may be puzzled as to its identity: is it a plant or a fungus? Indeed, the waxy, candlelike stems seem most unplantlike. Fringed pinesap, like a number of other members of the Heath family, has adapted to living on shaded forest floors by becoming saprophytic (absorbing its nutrients from decaying matter in the humus) and thus lacks green leaves and chlorophyll. All that remains of the foliage is thick, scalelike appendages hugging the stem and having little apparent usefulness. Reddish or golden brown, the stems may reach 12 inches; at other times they scarcely break the surface, so that only the watchful will detect the little reddish or yellow bulges in the bed of needles. In late summer, the stems turn blackish. Fairly common on Mt. Constitution, e.g., at Cold Springs, Mountain Lake, etc.; widespread, although scattered, elsewhere in the archipelago.

PUGET BUTTERWEED *(Senecio macounii)*. Splashes of buttery yellow along the road up to and at the summit are provided by this handsome wildflower. Easily identified by the bright yellow flowers present from July to September, clustered atop a single stem; below are long, slender, stalked leaves. Common on Mt. Constitution in open, rocky sites, occurring also (rarely) on Blakely Island. A notable regional endemic, found strictly from southwestern British Columbia to the Willamette Valley, Oregon. Rayless mountain butterweed *(S. indecorus)* grows on low swampy ground near Summit Lake and reputedly at the Beaverton Marsh, San Juan Island. It has long, irregularly lobed leaves, yellow flowers that lack petals, and up to 2-foot stems.

LOW MOUNTAIN GOLDENROD *(Solidago spathulata* var. *neomexicana)*. An easily overlooked species because it grows in with Puget butterweed and, like it, provides brilliant yellow accents to Mt. Constitution's rocky slopes well into September, long after most other wildflowers have withered. The flowers are smaller than those of Puget butterweed and, more importantly, are scattered along the stem, unlike the terminal cluster of Puget butterweed. Long, spathulate (i.e., spoon shaped, hence the specific name) leaves, forming a basal rosette, are also a good field mark. Apparently restricted to Mt. Constitution in the San Juans and, again, a more typical species of foothills to mid-montane elevations of the Cascades and Olympics.

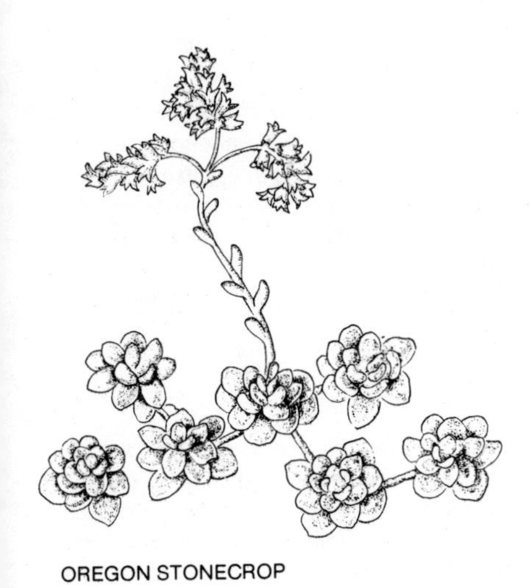

OREGON STONECROP

FRINGED PINESAP

PUGET BUTTERWEED

LOW MOUNTAIN GOLDENROD

Red

RED COLUMBINE *(Aquilegia formosa).* Certainly one of the most dazzling of wildflowers with its intense scarlet spurs and petals. Hardly less astounding are the bright yellow "honeycomb" heads located below, the sweet honey glands within attracting eager hummingbirds. When not in bloom the delicate, thin leaves are distinctive: each is divided into three grayish green leaflets, each of these in turn trilobed. Grows on steep, moist, rocky slopes near Mt. Constitution's summit and at several adjacent locations. Most common on Orcas but also on San Juan and other islands.

Pink

SPREADING DOGBANE *(Apocynum androsaemifolium* var. *pumilum).* Droopy, dark green leaves hanging oppositely from the stem are the best field mark. From May to July, the tiny white to pinkish bell flowers hang in open clusters from the branch tips. Milky juice oozes out if the stems (to 2 feet tall) are broken. Indian hemp, a close relative, was used by native Americans, who shredded the sinewy stems into fishing lines and ropes. Spreading dogbane is mostly restricted to Mt. Constitution in our area, occurring on dry, rocky places; a small disjunct population occurs near the Mar Vista Resort, San Juan Island.

Blue

SHOWY POLEMONIUM *(Polemonium pulcherrimum* var. *pulcherrimum).* During April and May, the naturalist on Mt. Constitution may be surprised to find a cluster of pale blue flowers already in full bloom while other species have barely begun their rejuvenation. Although represented only by scattered individual plants, the heavenly blooms make this wildflower worthy of the search. Showy is an apt description for the blue cups, each usually about ½ inch wide and the whole mass forming a dense cluster. Also called Jacob's ladder, a name based on the opposite, ladderlike arrangement of the leaves, which have a rank odor. Found on open, rocky slopes; in addition to the Mt. Constitution population, this species is also present on Entrance Mountain, Orcas Island, and at sea level on Iceberg Point, Lopez Island. Also on Mt. Erie, Fidalgo Island, to the east.

BLUEBELLS OF SCOTLAND *(Campanula rotundifolia).* Strikingly beautiful when the flowers appear in June, this species is another treat for those who traverse the steep, rocky slopes. One would be hard pressed to find a wildflower more magnificent than this one, with its unmistakable sky-blue to purplish cups. The 3/4–1-inch-long blooms seem too big for the thin, vining stems, which revel in creeping up the steepest slopes available. Small, linear leaves are scattered along the stems, although those of younger plants are ovate to irregularly lobed. Grows vigorously at certain places on Mt. Constitution and nearby, and also occurs sparingly elsewhere in the islands. The nearly vertical slopes at Chadwick Hill, Lopez Island, host a very substantial colony. *Campanula*, a genus name with a pleasing euphony, is Latin for little bell.

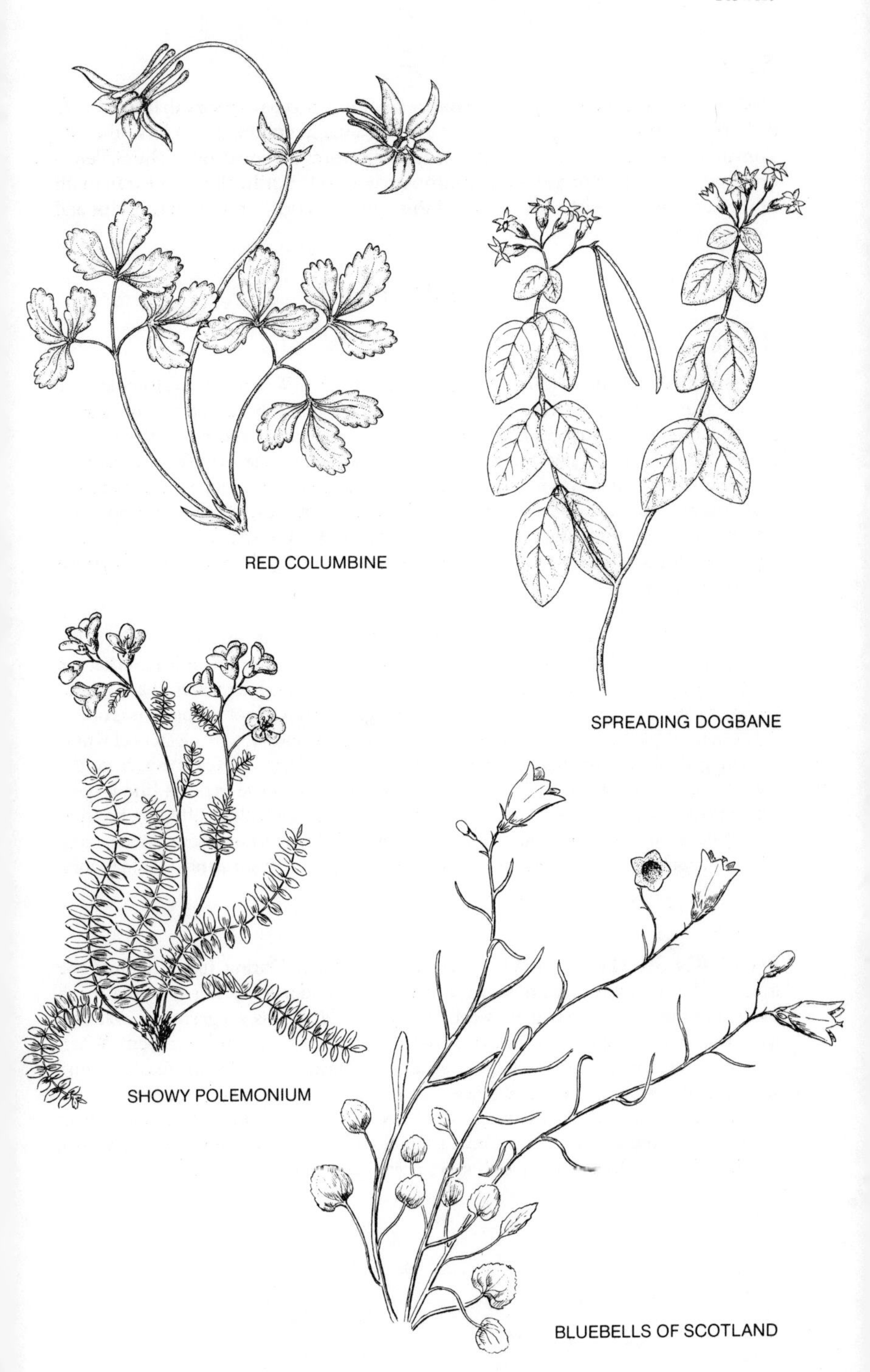
RED COLUMBINE
SPREADING DOGBANE
SHOWY POLEMONIUM
BLUEBELLS OF SCOTLAND

Brown

DWARF MISTLETOE *(Arceuthobium tsugense)*. A peculiar species that makes life difficult for the trees it parasitizes. Although usually attacking hemlock, the Mt. Constitution population lives on Lodgepole pine, causing a stunted form. The golden to brownish stems are thin, jointed, and inconspicuous—1–3 inches long. The trail south of the summit is a reliable place to find this species; watch for it on branch tips and young limbs.

SHRUBS

White

HAIRY MANZANITA *(Arctostaphylos columbiana)*. Where the Lodgepole pine woods open up south of the summit, low mats of this shrub may be found. Woody, crooked, and deeply rust colored, the limbs, plus the urn-shaped white flowers and ovate, evergreen leaves, give the impression of a dwarf madrona, to which manzanita is closely related. Watch also for the characteristic olive-greenish leaf color and densely hairy twigs. The May flowers are followed by blackish berries, formerly an important food source of various indigenous peoples. Plants on Mt. Constitution, nowhere taller than 2 feet, are smaller than normal size. Also occurs on rugged, open ridges of Cypress Island eastward.

BOG LABRADOR-TEA *(Ledum groenlandicum)*. Larger but similar in foliage to the preceding species, Labrador-tea has broader, somewhat lighter green leaves that are reddish-woolly below, and it is bigger—to 3 feet in height. Entrancing the wayfarer before he has even set foot on the bog, the unforgettable aroma of the blooms is one of nature's most pleasant scents. Elegant well describes the showy, tight clusters of white, star-shaped flowers themselves, present from June to August. Tea made from the leaves is very strong and should be used with discretion; some persons find it quite pleasant-tasting, while others do not. In either case, it is reputedly an effective laxative. Also present on the sphagnum bogs at Summit Lake and elsewhere. Much more common in the far north, where it is often the most abundant shrub on upland tundra.

Purple

WESTERN SWAMP-LAUREL *(Kalmia occidentalis)*. During June and July the lovely reddish purple flowers serve notice that Labrador-tea is not the only shrub present in the sphagnum bog. However, the attractive blooms, suggestive of azalea or rhododendron, are small (to ½ inch or so wide) and one must watch for them. When not in bloom, plants are inconspicuous among thickets of Labrador-tea. Twisting stems, either creeping or erect, bear leathery, evergreen leaves that are curled downward on the margin; also white-woolly below. Found on the sphagnum islets in Summit Lake and at a few other Orcas Island locations; also at the Beaverton Marsh, San Juan Island. The genus name honors Peter Kalm, a student of Linnaeus.

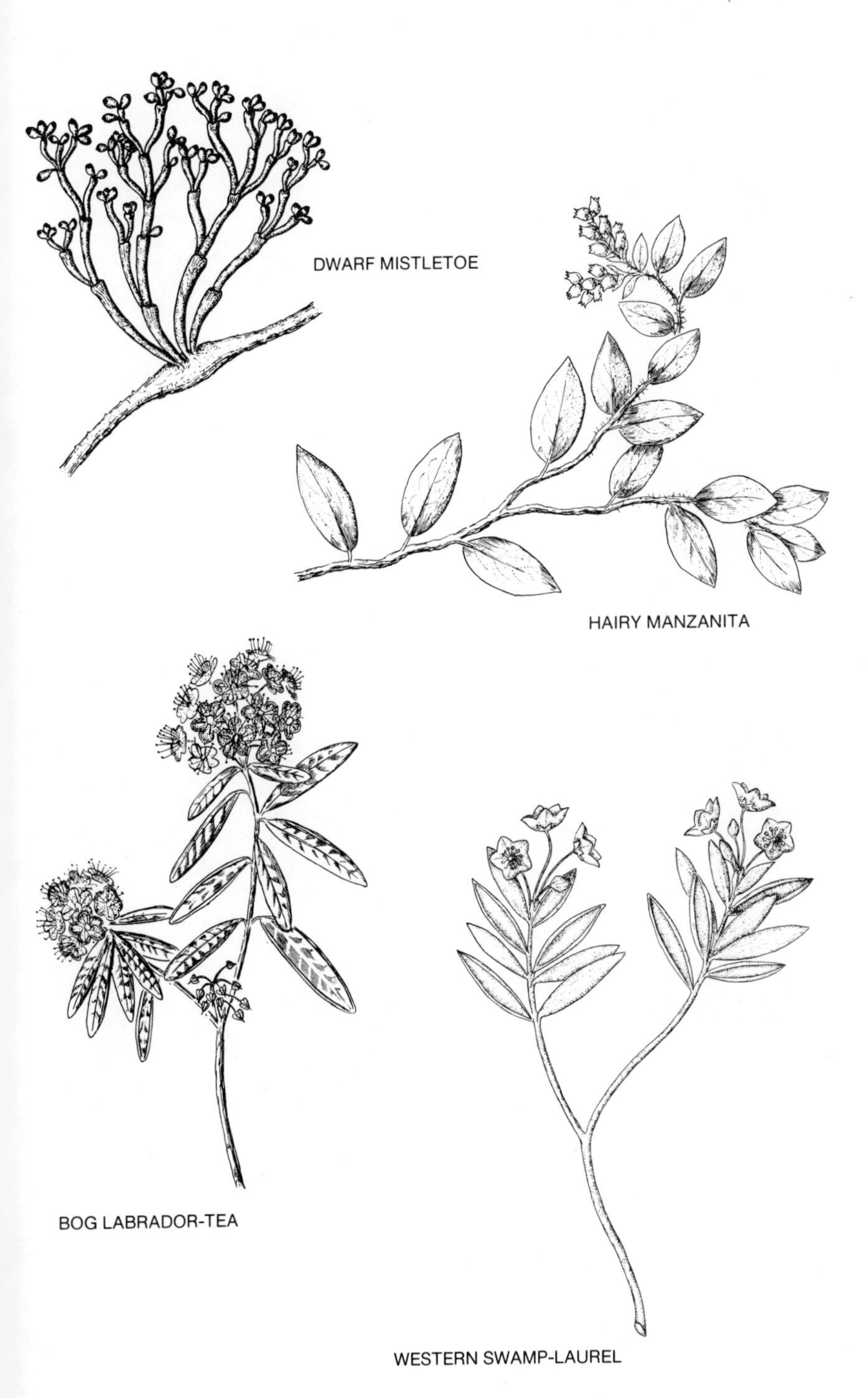
DWARF MISTLETOE
HAIRY MANZANITA
BOG LABRADOR-TEA
WESTERN SWAMP-LAUREL

—APPENDIX A—
GLOSSARY

achene: A small, dry fruit that does not split open at maturity.

adventive: A non-native plant, generally weedy and of Eurasian origin.

anther: The pollen-bearing part of a stamen.

axillary: Located or arising between the stem of a leaf and the stem of the plant.

basal sheath: A tubular structure surrounding the lower part of a stem.

bract: A modified or specialized leaf, from the axil of which arises an inflorescence or flower.

bracteole: A small modified leaf, usually found in association with an inflorescence; a small bract.

calyx: Collectively, all of the sepals.

corolla: Collectively, all of the petals.

endemic: A species restricted to only one particular area or geographical region.

ephemeral: A plant that is present for a brief period, generally in early spring.

epiphyte: A plant that grows upon another plant, using it for anchorage but not obtaining nutrients from it.

filament: The stalk of a stamen.

herbaceous: Non-woody; dying back to the ground at the end of the growing season.

hybridize: The cross-breeding of two different species.

inflorescence: The collective arrangement of flowers on the stem.

involucre: A set of small, leaflike appendages found at the base of an inflorescence.

lobe: The projecting part of a structure.

ovary: The seed-producing portion of a flower.

panicle: A branching flower stem.

pedicel: The stalk of a single flower in an inflorescence.

perfoliate: A leaf with the basal portions joined around the stem.

petal: The expanded, often brightly colored part of a flower.

pinnate: Divided to the axil of a leaf (compound), such that the leaflets are arranged on each side of a common stalk, as with a feather.

pistil: The female organ of a flower, consisting of ovary, style, and stigma.

raceme: A long unbranching inflorescence.

rhizome: A creeping underground stem from which individual, upright stems arise.

rosette: A cluster or whorl of leaves, usually at the base of a plant.

sepals: The outermost parts of a flower, just below the petals, usually greenish and leafy.

spike: A tall inflorescence, with the flowers clustered and attached directly to the stem.

filament: The stalk of a stamen.

herbaceous: Non-woody; dying back to the ground at the end of the growing season.

hybridize: The cross-breeding of two different species.

inflorescence: The collective arrangement of flowers on the stem.

involucre: A set of small, leaflike appendages found at the base of an inflorescence.

lobe: The projecting part of a structure.

ovary: The seed-producing portion of a flower.

panicle: A branching flower stem.

pedicel: The stalk of a single flower in an inflorescence.

perfoliate: A leaf with the basal portions joined around the stem.

petal: The expanded, often brightly colored part of a flower.

pinnate: Divided to the axil of a leaf (compound), such that the leaflets are arranged on each side of a common stalk, as with a feather.

pistil: The female organ of a flower, consisting of ovary, style, and stigma.

raceme: A long unbranching inflorescence.

rhizome: A creeping underground stem from which individual, upright stems arise.

rosette: A cluster or whorl of leaves, usually at the base of a plant.

sepals: The outermost parts of a flower, just below the petals, usually greenish and leafy.

spike: A tall inflorescence, with the flowers clustered and attached directly to the stem.

stamen: The male pollen-producing organ of the flower, including the anthers and often a filament.

staminoidia: Modified stamens.

stigma: The female portion of the flower that receives pollen.

style: The elongate stalk that joins the stigma to the ovary.

umbel: A flower cluster in which the flower stalks (pedicels) radiate from the same points, often forming a symmetric, flat-topped or convex bloom.

whorl: A ring of three or more similar structures (usually leaves) surrounding a stem or radiating from a common point.

APPENDIX B

VASCULAR PLANTS OF SAN JUAN COUNTY

At the time of printing, 829 vascular plant species, belonging to 388 genera and 98 families, have been recorded as naturalized or native to San Juan County. 283 of the taxa, or roughly 33 percent, are introduced, the majority of them European adventives.

Compiling a vascular plant list for an area with the size and inaccessibility of San Juan County is, to say the least, an enormous task. In fact, the list section itself represents many hours of field study, of herbarium visits, of library research and personal communications, plus much more. Several years were required, and the period was marked by moments of joy and, occasionally, of frustration. Discovering quite by accident that, despite the disturbance of all traditional sites, a single population of the rare Golden Indian paintbrush *(Castilleja levisecta)* is lingering on San Juan Island, and that a completely unexpected voucher of Golden-eyed grass *(Sisyrinchium californicum)* is among the hundreds of thousands of specimens at the University of Washington herbarium, are, for example, two of the more exciting moments we have had. On the other hand, sifting through scores of old publications, personal reports, and specimens, only to discover that many were incomplete or in error, was at times a disappointment. Nonetheless, the compilation was a rich learning experience, and one that leaves us with many good memories.

A number of site-specific reports have been made of single islands or plots, and there has historically been no lack of interest in the San Juans: several famous Northwest botanists collected in the islands, and numerous direct and indirect references have been made to San Juan flora in publications as varied as the *San Juan County Almanac* and Piper's *Flora of Washington.* Nonetheless, the understanding of San Juan flora appears scant in comparison to that of such places as the Olympic Peninsula or Vancouver Island. Indeed, many more species will eventually be located in the San Juans, and a more complete understanding of island vegetation will be achieved, as interest and coverage increase.

San Juan floral composition bears close resemblance to that of other parts of the rain shadow region west of the Cascades, sharing the general characteristics of playing host to a number of standard east of the Cascades, dry belt taxa and featuring a declining influence by several species of the wet coniferous forest. However, the archipelago's position and geologic history have caused some fascinating variations on the rain shadow theme. Thus, in addition to having several taxa more commonly found east of the Cascades, the archipelago in places reveals affinities with the flora of the outer coast, notably of species not occurring farther north than southern Oregon except for disjunct populations in the San Juans and adjacent Vancouver Island. More prominent is the existence of a relic late glacial flora on the summit of Mt. Constitution, where several markedly subalpine and northern species linger, a few of them apparently near the southern and low elevation maximums of their range. The presence of de-alpinized taxa at a few sea-level sites is also diagnostic. Finally, the absence or scarcity of many typical species of the Western hemlock zone reinforces the island isolation theme.

Many of the more strikingly absent or nearly absent taxa are denizens of the moist, shady woodland and adjacent stream valley, including:

Vine maple *(Acer circinatum)*
Devil's club *(Oplopanax horridum)*
Pacific rhododendron *(Rhododendron macrophyllum)*
Vanilla-leaf *(Achlys triphylla)*
California hazelnut *(Corylus cornuta)*
Slender waterleaf *(Hydrophyllum tenuipes)*
Western trillium *(Trillium ovatum)*
Western baneberry *(Actaea rubra)*
Youth-on-age *(Tolmiea menziesii)*
Deer fern *(Blechnum spicant)*
Oak fern *(Gymnocarpium dryopteris)*
Stream violet *(Viola glabella)*
Evergreen huckleberry *(Vaccinium ovatum)*
Indian-plum *(Osmaronia cerasiformis)*
Western bleeding heart *(Dicentra formosa)*
Giant Solomon's seal *(Smilacina racemosa)*
Cascara *(Rhamnus purshiana)*

Contrastingly, a number of interesting species that are more common east of the Cascades have been recorded in the San Juans, including:

Rocky Mountain woodsia *(Woodsia scopulina)*
Rocky Mountain juniper *(Juniperus scopulorum)*
Sand fringepod *(Thysanocarpus curvipes)*
Contorted-pod evening-primrose *(Oenothera contorta)*
Prickly pear cactus *(Opuntia fragilis)*
Soopolallie *(Shepherdia canadensis)*
Large-flowered collomia *(Collomia grandiflora)*
Thread-leaved phacelia *(Phacelia linearis)*
Slender plagiobothrys *(Plagiobothrys tenellus)*
Rosy owl-clover *(Orthocarpus bracteosus)*
Small fleabane *(Erigeron speciosus)*
Lemmon's needlegrass *(Stipa lemmonii* v. *lemmonii)*
Pine bluegrass *(Poa scabrella)*

There are several other species and subspecies present in the archipelago that are generally more often associated with the dry east side of the Cascades.

Several chiefly outer-coast species are also found in the San Juans, where access to parent populations via the Strait of Juan de Fuca and the availability of appropriate habitat are the probable causes for their presence. In this vein, it is notable that they are mostly found on the southern edge of the archipelago, particularly in the sand-dune areas of Cattle Point and adjacent southern San Juan Island, and on southern Lopez Island. This group includes: Beach morning glory *(Convolvulus soldanella),* Coastal strawberry *(Fragaria chiloensis),* Lesueur's salt-rush *(Juncus lesueurii),* and Sand-dune sedge *(Carex pansa).* Many more species that occur sparingly on strand habitats of Puget Sound but commonly along the outer coast are also present.

More difficult to understand is the occurrence of several species, mostly confined to coastal California and southern Oregon, disjunct in the San Juans and adjacent rain shadow areas, including California buttercup *(Ranunculus californicus),* Greene's rein-orchid *(Habenaria greenei),* Coast microseris *(Microseris bigelovii),* Sharp-fruited peppergrass *(Lepidium oxycarpum),* Slender woolly-heads *(Psilocarpus tenellus),* and Pygmy tillaea *(Tillaea erecta),* the latter three originally found in the islands by Dr. Adolf Ceska of Victoria. This group is noteworthy in its affiliation with vernal pool or moist meadow habitats.

Although southerly, outer coastal, and easterly influences are interesting characteristics of San Juan flora, no element is more distinctive than the presence of northerly and middle-upper elevation Cascade-Olympic mountain species on Mt. Constitution. Some of the more interesting taxa include:

Alpine lady fern *(Athyrium distentifolium)*
Mountain licorice fern *(Polypodium amorphum)*
Long-stalked starwort *(Stellaria longipes* var. *altocaulis)*
Oregon stonecrop *(Sedum oreganum)*
Matted saxifrage *(Saxifraga bronchialis* var. *austromontana)*
Alpine willow-herb *(Epilobium alpinum* var. *alpinum)*
Mountain mare's tail *(Hippuris montana)*
Showy polemonium *(Polemonium pulcherrimum* var. *pulcherrimum)*
Rosy pussytoes *(Antennaria microphylla)*
Aleutian wormwood *(Artemisia tilesii* var. *unalaschcensis)*
Arctic aster *(Aster sibiricus* var. *meritus)*
Dwarf mountain daisy *(Erigeron compositus* var. *glabratus)*
Rayless mountain butterweed *(Senecio indecorus)*
Few-flowered sedge *(Carex pauciflora)*
Idaho fescue *(Festuca idahoensis* var. *idahoensis)*

Although the phenomenon of de-alpinization is mostly confined to Mt. Constitution, in a few places it occurs at sea level in the islands. Our de-alpinized sea-level taxa include Menzies' silene *(Silene menziesii* var. *menziesii),* Matted saxifrage, and two dramatic examples at Iceberg Point, Lopez Island: Slender crazyweed *(Oxytropis campestris* var. *gracilis)* and Showy polemonium.

The earliest known collections of San Juan County plants were those of Captain Archibald Menzies, who stopped on Orcas Island in early June 1792 during his famous voyage with Captain George Vancouver aboard the *Discovery.* Menzies, one of the Northwest's most famous botanists, noted in his memoirs the "rugged and cliffy" shores of Orcas and the interesting flora thereon; the most notable of his collections was Kamchatka fritillary *(Fritillaria camschatcensis),* a rarity in Washington and perhaps no longer existent in the archipelago. Dr. David E. Lyall, another prominent figure in early Northwest botany, surveyed the flora of Lopez, San Juan, and Orcas Islands during 1858, finding several outstanding rarities that have not been relocated. Several of his collections apparently remain at the Gray Herbarium in Cambridge, Massachusetts, but, alas, our correspondence with personnel there reveals that several of the vouchers cited by C. V. Piper (1906) have now been lost or misplaced. Lyall's most notable finds include Footsteps-of-spring *(Sanicula arctopoides),* in the Washington part of its range restricted to one location, and White meconella *(Meconella oregana),* a small white poppy lingering only on Whidbey Island in the Washington part of its range. Following Lyall was L.F. Henderson, who, collecting plants statewide for the World's Fair in 1892, passed through the islands making several notable collections on Mt. Constitution.

A number of collectors visited the archipelago in 1904, including A. S. Foster and A. S. Pope (on Orcas and San Juan Islands chiefly) and W. H. Lawrence (Stuart Island). In 1908 F. H. Frye collected several notable species on San Juan Island, followed by A. R. Roos there in 1910. Morton E. Peck did a June-July stint in 1923 on the larger islands, making a number of important collections. In 1940 J. S. Martin covered Shaw and Sucia Islands, two islands mostly ignored previously.

During the modern period important vouchers came from Lenora M. Sunquist (1964), Dr. Melinda Denton (mid-1970s), currently professor of botany at the University of Washington and Peter Rapp (1980-81).

Aside from vouchers, several key observers have given important information regarding San Juan flora through written or verbal channels. Betty Higgenbotham, an excellent resident botanist, has written a number of articles and taught classes on San Juan flora; several of her records are cited in this list. William Baker is another excellent resident botanist who has provided a number of good records; he has botanized the archipelago since 1947. Several more persons, some of them affiliated with The Nature Conservancy, have added to our knowledge of the islands' flora by publishing ecological reports on Yellow, Sentinel, Goose, and other islands; Charles Eaton, Dr. Eugene Kozloff, Peter Dunwiddie, and Peter Rapp are among those involved. The Washington Natural Heritage Program, monitoring rare taxa statewide, has been helpful in giving exact whereabouts of several species, and one of their botanists, Linda Kunze, has found a number of excellent species recently in the county's wetlands. Dr. Adolf Ceska, from the Victoria Provincial Museum, is responsible for having found four of the county's rarest plants, and another Canadian, Harvey Janzsen, has added several excellent records. Other individuals have alerted the authors to other rare species or have written material concerning San Juan vegetation; their names can be found in the acknowledgments or in the bibliography.

The sequence of families follows that of *Vascular Plants of the Pacific Northwest* by C. L. Hitchcock and Arthur Cronquist (1973); with a few exceptions, nomenclature also follows that work. General information regarding range, distribution, and habitat are given for each species, occasionally augmented by statements concerning endemism, intergradation, and variations involved. No description is given for taxonomic characteristics; consult Hitchcock (1973) or any of a number of other publications for that information. Less common species tend to have longer descriptions. An asterisk preceding a scientific name indicates that the species is non-native, most of our species of this type having been introduced from Eurasia. The name or names following each Latin name and preceding the common name are those of the one or more botanists who originally described the species. These names allow interested readers to look up those original descriptions to see if they find them taxonomically acceptable. Abbreviations for botanists' names are listed in *Manual of Cultivated Plants* by L. H. Bailey.

Vouchers are listed for rare and interesting species. They are located in herberia at the University of Washington (WTU), Western Washington State University (WWB), Washington State University (WS), University of Puget Sound (UPS), the Victoria Provincial Museum (V), and the Friday Harbor Laboratories (FH). In addition, some vouchers are present in the personal collections of Harvey Janzsen (HJ) and in those of the authors (SA) and (FS). Vouchers are also present for the most common plants at WTU and FH.

We encourage readers to contact us regarding any questions, and we would especially appreciate any information concerning new species for the county or range expansions.

Fred Sharpe
280 Sharpe Road
Anacortes, Washington 98221

Scott Atkinson
18944 40th Place NE
Seattle, Washington 98155

ANNOTATED CHECKLIST

Lycopodiaceae

Lycopodium clavatum L. Stag's horn moss—Rare; deep coniferous forest. Mountain Lake and Twin Lakes trail, Orcas Island. (WWB)

Selaginellaceae

Selaginella wallacei Heiron. Wallace's selaginella—Abundant; mossy bedrock outcrops.

Isoetaceae

Isoetes echinospora Dur. Bristle-like quillwort—Aquatic; reported from Spencer Lake, Blakely Island. (J.R. Slater)

Isoetes nuttallii A. Br. Nuttall's quillwort—Rare; vernal pools; collected by Dr. Adolf Ceska near Cattle Point in May, 1983. (V)

Equisetaceae

Equisetum arvense L. Field horsetail—Common; margins, shady woods, weedy.

Equisetum fluviatile L. Water horsetail—Rare; present on the margins of a few lakes on Orcas Island. (UPS)

Equisetum hyemale L. var. *affine* Dutch horsetail—Locally abundant; glacial outwash seeps, interdune troughs, moist stream valleys.

Equisetum telmateia Ehrh. var. *braunii* Giant horsetail—Common; damp places.

Ophioglossaceae

Botrychium multifidum (Gmel.) Trevis. Leathery grape-fern—Moist meadows to sphagnum bogs; found on Lopez, Shaw, Orcas, San Juan, and Sucia islands.

Botrychium virginianum (L.) Swartz. Virginia grape-fern—Rare; stream valleys near Killebrew Lake, Orcas Island.

Ophioglossum vulgatum L. Adder's tongue—Rare; sphagnum bogs, west lobe of Orcas Island. (FS)

Polypodiaceae

Adiantum pedatum L. Maidenhair fern—Locally common, shady stream valleys and waterfalls, Mt. Constitution and vicinity; scarce elsewhere.

Aspidotis densa (Brackenr.) Lellinger Indian's dream—Locally abundant, open meadows, Mt. Constitution; a reliable indicator of serpentine.

Asplenium trichomanes L. ssp. *trichomanes* Maidenhair spleenwort—Rare; shaded, wet rock crevices; Mt. Constitution, Mt. Dallas, and Blakely Island. Ssp. *quadrivalens* has not been documented but should be watched for on limestone outcrops in the San Juans. (WTU)

Athyrium distentifolium Tausch Alpine lady fern—Rare; wet rock seeps, upper elevations, Mt. Constitution. Originally collected by Pope in August, 1904. (WTU)

Athyrium filix-femina (L.) Roth. Lady fern—Common; moist woods, fens, and open wet waste places.

Blechnum spicant (L.) Roth. Deer fern—Deep, moist coniferous woods; uncommon but widespread, being most common on Blakely and Orcas islands.

Cryptogramma acrostichoides R. Brown Parsley fern—Locally common, open rocky meadows, Mt. Constitution; scarce elsewhere. Listed in Hitchcock as *C. crispa.*

Cystopteris fragilis (L.) Bernh. Bladder fern—Shaded streambanks and steep, moist rocky crevices on Mt. Constitution; locally common; rare, San Juan Island.

Dryopteris expansa (Presl.) Fraser-Jenkens and Germy Mountain wood fern—Deep, moist coniferous woods; fairly common, often growing on rotting stumps. Listed in Hitchcock as *D. austriaca.*

Gymnocarpium dryopteris (L.) Newm. Oak fern—Deep, moist coniferous woods; reported by G. Sharpe from Patos Island.

Pityrogramma triangularis (Kaulf.) Maxon. var. *triangularis* Gold-back fern—Shaded to fairly exposed rock crevices; scattered but fairly common throughout.

Polypodium amorphum Suksd. Mountain licorice fern—Steep, mossy rock slopes; fairly common.

Polypodium glycyrrhiza D. C. Eat. Licorice fern—Common; mossy rock slopes to forest margins; generally not epiphytic on *Acer macrophyllum* in our area.

Polystichum imbricans (D.C. Eat.) D.H. Wagner Imbricate sword fern—Rocky, sunny slopes, mostly at upper elevations; occasional. (UPS)

Polystichum munitum (Kaulf.) Presl. var. *munitum* Sword fern—Common to abundant; moist woods.

Pteridium aquilinum (L.) Kuhn. var. *pubescens* Bracken fern—Abundant; forest margins to grassy meadows.

Woodsia scopulina D.C. Eat. Rocky Mountain woodsia—Open to partially shaded rock crevices, locally common on Mt. Constitution; scarce elsewhere.

Taxaceae

Taxus brevifolia Nutt. Pacific yew—Coniferous forest understory, often where steep; fairly common.

Cupressaceae

Juniperus scopulorum Sarg. Rocky Mountain juniper—Open rocky slopes near salt water; common.

Thuja plicata Donn Western red-cedar—Valley bottoms, north-facing slopes, usually where moist; common.

Pinaceae

Abies grandis (Dougl.) Forbes Grand fir—Variable in lowland habitats; common.

Picea sitchensis (Bong.) Carr. Sitka spruce—Swampy soils near fresh water, less frequently shorelines and moist mixed woodlands; fairly common.

Pinus contorta Dougl. var. *latifolia* Lodgepole pine—Common to abundant; dry, sandy areas to moist woods and boggy lake shores.

Pinus monticola Dougl. Western white pine—Local on Mt. Constitution, where mostly represented by young trees; rare elsewhere.

Pseudotsuga menziesii (Mirbel) Franco. var. *menziesii* Douglas-fir—Abundant throughout.

Tsuga heterophylla (Raf.) Sarg. Western hemlock—Dominant in denser coniferous forests, north-facing slopes, etc.; largest concentrations on Blakely and Orcas islands.

Salicaceae

Populus tremuloides Michx. Quaking aspen—Widespread and locally common on larger islands; riparian habitats and low glacial substraits.

Populus trichocarpa T.&.G. Black cottonwood—Riparian; rather rare, mostly on San Juan Island. (WTU)

Salix hookeriana Barratt Hooker's willow—Sand dunes, rocky bluffs near salt water, and other maritime habitats; fairly common throughout.

Salix lasiandra Benth. var. *lasiandra* Pacific willow—Riparian to aquatic; common.

Salix scouleriana Barratt Scouler's willow—Common to abundant in shady woods, margins, open meadows, and lake shores.

Salix sitchensis Sanson Sitka willow—Lake edges; common on larger islands.

Betulaceae

Alnus rubra Bong. Red alder—Disturbed open places, margins, and wet places; common.

Alnus sinuata (Regel) Rydb. Sitka alder—Local on the summit of Mt. Constitution; open rocky slopes. Disjunct at Spencer Spit State Park, Lopez Island. (WTU)

Betula papyrifera Marsh var. *commutata* Paper birch—Stream valleys, riparian habitats; occasional.

Corylus cornuta Marsh var. *californica* California hazelnut—Introduced but apparently rare as a native, as in the Wasp Island group; moist woods.

Fagaceae

Quercus garryana Dougl. Garry oak—Dry, rocky sites, often near salt water; most common on San Juan and Orcas islands but found throughout.

Urticaceae

Urtica dioica L. ssp. *gracilis* var. *lyallii* Stinging nettle—Low, wet places and in disturbed woodlands, often where soil is fertile; abundant.

**Urtica urens* L. Dog nettle—Garden weed; occasional, Friday Harbor. European.

Loranthaceae

Arceuthobium tsugense Hawksw. & Weins Dwarf mistletoe—Locally common on Mt. Constitution, where attacking *Pinus contorta;* also reported from San Juan Island by Baker and others. (WTU)

Polygonaceae

Polygonum amphibium L. Water smartweed—Common aquatic.
**Polygonum aviculare* L. Doorweed—Common European weed.
Polygonum coccineum Muhl. Water smartweed—Riparian to semi-aquatic; locally common on San Juan and Orcas islands.
**Polygonum convolvulus* L. Black bindweed—European weed; occasional in gardens.
Polygonum cuspidatum Sieb. & Zucc. Japanese knotweed—Asiatic; cultivated and occasionally escaping, as on San Juan Island.
Polygonum fowleri Robins. Fowler's knotweed—Strand, salt marshes; fairly common throughout.
Polygonum nuttallii Small Nuttall's knotweed—Dry, rocky prairies; chiefly Mt. Constitution. (WTU)
Polygonum paronychia Cham. & Schlecht. Beach knotweed—Sand dunes; reported by Kozloff from South Beach, San Juan Island.
**Polygonum persicaria* L. Lady's thumb—Riparian, wet waste places, and gardens; occasional.
**Polygonum sachalinense* Schmidt Giant knotweed—Cultivated and ± escaping/spreading, mostly larger islands. Native of Asia.
Polygonum spergulariaeforme Meisn. Fall knotweed—Locally abundant; dry rock outcrops and meadows.
**Rumex acetosella* L. Sheep sorrel—Weedy sites to rocky outcrops; abundant.
**Rumex conglomeratus* Murr. Clustered dock—Occasional weed, larger islands.
**Rumex crispus* L. Curly-leaved dock—Lake edges, meadows, shorelines, and weedy places; common throughout; European.
Rumex maritimus L. Seaside dock—Brackish marshes; occasional.
**Rumex obtusifolius* L. Bitter dock—Common weed, European.
Rumex occidentalis Wats. var. *procerus* Western dock—Salt marshes and strand; widespread but most common from Roche Harbor south to Cattle Point, San Juan Island.
Rumex salicifolius Weinm. ssp. *salicifolius* Willow dock—Wet waste places to strand; fairly common.

Chenopodiaceae

Atriplex patula (L.) Gray var. *hastata* Common orache—Abundant, maritime.
**Chenopodium album* L. Lambsquarter—Common weed.
**Chenopodium murale* L. Sowbane—Rare; disturbed places, Friday Harbor.
**Chenopodium rubrum* L. Red goosefoot—Salt marshes; widespread but uncommon.
**Salicornia europaea* L. European glasswort—Local, salt marshes on Lopez Island. (WWB)
Salicornia virginica L. Pickleweed—Abundant; the dominant species of salt marshes.

Amaranthaceae

**Amaranthus graecizans* L. Prostrate pigweed—Local, sand dunes at Cattle Point.
**Amaranthus powellii* Wats. Powell's pigweed—Occasional European weed of gardens.
**Amaranthus retroflexus* L. Rough pigweed—Waste places; reported by Baker and others.

Nyctaginaceae

Abronia latifolia Eschsch. Yellow sand-verbena—Strand; locally abundant, mostly on San Juan and Lopez islands.

Portulacaceae

Calandrinia ciliata (R. & P.) DC. Desert rockpurslane—Uncommon; open meadows; most common on San Juan Island. (WTU)
Montia dichotoma (Nutt.) Howell Dwarf montia—Moist meadows and seeps; occasional; chiefly on Mt. Constitution and Mt. Dallas–Cady Mountain. (SA)
Montia fontana L. var. *tenerrima* Water chickweed—Moist meadows and wet coniferous forest margins; widespread but inconspicuous. (FS)
Montia linearis (Dougl.) Greene Narrow-leaved montia—Dry to moist meadows and occasionally weedy; fairly common throughout.

Montia parvifolia (Moc.) Greene Streambank spring beauty—Streambanks and moist seeps to wet meadows; widespread. Var. *flagellaris* is our common form, var. *parvifolia* occasional; some of our material appears to be intermediate.
Montia perfoliata (Donn) Howell Miner's lettuce—Dry coniferous woods to open meadows; common. Var. *perfoliata* is the common woodland form; var. *glauca* is fairly common on open meadows to strand.
Montia sibirica (L.) Howell var. *sibirica* Siberian miner's lettuce—Moist coniferous to deciduous woodlands and margins; common.
Montia spathulata (Dougl.) Howell Pale montia—Dunes and dry meadows; locally common on San Juan Island and occasional elsewhere.

Caryophyllaceae

Arenaria macrophylla Hook. Big-leaved sandwort—Moist coniferous forests and margins; fairly common on Orcas Island, especially on Mt. Constitution; less common on Mt. Dallas and elsewhere.
Arenaria paludicola (Robins.) Swamp sandwort—Shady, swampy places; rare; known from a single station on the north end of Stuart Island, where it grows with typical associates such as Common rush *(Juncus effusus* var. *effusus),* Golden-seeded sedge *(Carex aurea)* and Blue-eyed grass *(Sisyrinchium angustifolium).* Listed as endangered in Washington by the Washington Natural Heritage Program, 4 historic sites, all west of the Cascades, were known to host populations, but all were recently declared extirpated, rendering the species extinct from Washington. Thus the Stuart Island record is the first new one for Washington and the only active colony. Swamp sandwort occurs chiefly in n. California, and is absent from Oregon and British Columbia. (FS)
**Arenaria serpyllifolia* L. Thyme-leaved sandwort—Dry gravelly meadows, occasionally dry coniferous woods; common. European.
Arenaria stricta Michx. var. *puberulenta* Slender sandwort—Dry grassy meadows and rocky outcrops; uncommon but widespread, up to higher elevations on Mt. Constitution.
Cerastium arvense L. Field chickweed—Abundant, meadows.
Cerastium nutans Raf. Nodding chickweed—Dry meadows; occasional.
**Cerastium tomentosum* L. Dusty miller—Cultivated in rock gardens and often escaping/spreading.
**Cerastium viscosum* L. Sticky chickweed—Common European adventive.
**Cerastium vulgatum* L. Common chickweed—Abundant weed; lawns to meadows.
**Dianthus armeria* L. Deptford pink—Occasional European weed, mostly on meadows near salt water.
Honkenya peploides (L.) Ehrb. Honkenya—Locally common, strand on Lopez Island from Odlin County Park to Mackaye Harbor; rarely elsewhere, as on Sucia Island.
**Lychnis alba* Mill. White campion—Occasional weed, larger islands. European.
**Lychnis coronaria* (L.) Desr. Rose campion—Cultivated and widely escaping; although uncommon, becoming a nuisance in some areas; European.
Sagina crassicaulis Wats. Stick-stemmed pearlwort—Rocky coastal bluffs to steep glacial banks, usually where moist and near salt water; common, especially along the southern edge of the archipelago.
Sagina occidentalis Wats. Western pearlwort—Meadows or waste places; uncommon. (WTU)
**Sagina procumbens* L. Procumbent pearlwort—European; common weed where moist.
**Silene antirrhina* L. Sleepy catchfly—European; occasional weed of dry meadows or coniferous forest margins; most common in the northern part of the archipelago.
**Silene cucubalus* Wibel Bladder campion—Occasional; East Sound. European.
**Silene gallica* L. Small-flowered catchfly—Meadows and disturbed open places; common.
Silene menziesii Hook. var. *menziesii* Menzies' silene—Rare; coastal meadows on southern San Juan Island. (SA)
Silene scouleri Hook. var. *pacifica* Scouler's silene—Dry, open coniferous forests and meadows; occasional. (WWB)
**Spergula arvensis* L. Cornspurry—Occasional European weed of gardens.
Spergularia canadensis (Pers.) G. Don Canadian sandspurry—Common; salt marshes.

Spergularia macrotheca (Hornem.) Heynh. Beach sandspurry—Salt marshes to rocky coastal bluffs; common.

**Spergularia marina* (L.) Griseb. Saltmarsh sandspurry—Occasional; salt marshes; European.

**Spergularia rubra* (L.) Presl. Red sandspurry—Weedy sites to strand; common.

Stellaria calycantha (Ledeb.) Bong. Northern starwort—Wet places; widespread. The common form is var. *bongardiana*, with var. *sitchana* occasionally intermixed.

Stellaria crispa Cham. & Schlecht. Crisped starwort—Deep, moist woods, and streambanks; common.

**Stellaria graminea* L. Lesser starwort—Ditches or lawns, European weed; occasional.

Stellaria longipes Goldie var. *altocaulis* Long-stalked starwort—Rare; known from a single wet meadow station on Mt. Constitution.

**Stellaria media* (L.) Vill. Chickweed—Common European adventive.

Stellaria nitens Nutt. Shining chickweed—Gravelly meadows; occasional.

**Vaccaria segetalis* (Neck.) Garcke Cowcockle—European weed; rare; collected by Frye on San Juan Island in 1908. (WTU)

Nymphaeceae

Brasenia schreberi Gmel. Water shield—Occasional; aquatic of lakes on Orcas Island. (FS)

Nuphar polysepalum Engelm. Indian pond-lily—Abundant, fresh water.

**Nymphaea odorata* Ait. American water-lily—Occasional; lakes.

Ceratophyllaceae

Ceratophyllum demersum L. Coontail—Aquatic of larger lakes; occasional; most numerous on Orcas Island.

Ranunculaceae

Anemone lyallii Britt. Lyall's anemone—Damp stream valleys and moist woods; most common on Orcas Island; less common on San Juan and other islands.

Aquilegia formosa Fisch. Red columbine—Shaded rocky seeps; most common on Mt. Constitution but also present on San Juan and other islands.

**Aquilegia vulgaris* L. Garden columbine—Cultivated and escaping along roadsides and waste places, Friday Harbor and East Sound.

**Clematis vitalba* L. Traveller's joy—Introduced and becoming abundant on San Juan Island, as at Roche Harbor; also in Friday Harbor and East Sound.

Delphinium menziesii DC. var. *menziesii* Menzies' larkspur—Meadows; fairly common throughout.

Delphinium nuttallii Gray Nuttall's larkspur—Coastal meadows and bluffs; rare; collected near Olga, Orcas Island and on Brown Island near Friday Harbor. These are apparently the northernmost records for this species, restricted to southern Puget Sound and Grays Harbor County south into Oregon, and along the Columbia River. (WS) (UPS)

Myosurus minimus L. Least mouse-tail—Vernal pools; widespread but scattered, from American Camp north to Mt. Dallas, and on Stuart (Baker) and Matia islands. (WWB)

**Ranunculus acris* L. Tall buttercup—Occasional weed, ditches and damp meadows; European.

Ranunculus aquatilis L. White water buttercup—Fairly common aquatic in larger lakes; var. *hispidulus* is our more frequent form, var. *capillaceus* restricted to Orcas Island.

Ranunculus californicus Benth. California buttercup—Locally abundant, open meadows, Iceberg Point north to Davis Head, Lopez Island, and from Cattle Point to Mt. Dallas, San Juan Island, and on adjacent small islets. Virtually unknown in Washington except from the San Juan Islands, this buttercup was originally found in the San Juans by Dr. Melinda Denton in 1976. It usually occurs and hybridizes with *R. occidentalis* in our area.

Ranunculus cymbalaria Pursh Seaside buttercup—Rare; brackish marshes; collected by Bensen in June, 1923 at Cattle Point. (WTU)

Ranunculus flammula L. Creeping spearwort—Common, riparian to semi-aquatic throughout.

Ranunculus occidentalis Nutt. var. *occidentalis* Western buttercup—Meadows; abundant.

Ranunculus orthorynchus Hook. var. *orthorynchus* Straight-beaked buttercup—Swampy or marshy places; reported by Higinbotham from the Lakedale drainage area, San Juan Island.

**Ranunculus repens* L. var. *repens* Creeping buttercup—Common European species of lawns, ditches, etc.

**Ranunculus sardous* Crantz Hairy buttercup—Well-established around Lopez Village, Lopez Island; apparently this is the first Washington record for this European adventive.

Ranunculus sceleratus L. var. *multifidus* Celery-leaved buttercup—Salt or brackish marshes; occasional in the northern part of the archipelago. (WTU)

Ranunculus uncinatus D. Don Little buttercup—Common; damp woods to open marshy areas. Var. *parviflorus* is widespread, var. *uncinatus* occasional.

Berberidaceae

Berberis aquifolium Pursh Tall Oregon-grape—Common, margins and open transitional forest.

Berberis nervosa Pursh Low Oregon-grape—Abundant; shady woodlands.

Papaveraceae

**Eschscholzia californica* Cham. California poppy—Locally abundant, meadows to roadside ditches; introduced and escaping widely, especially on San Juan Island.

Meconella oregana Nutt. White meconella—Rare; moist meadows; collected by David E. Lyall on Orcas Island in 1858. Rare in Washington. (Piper, 1906)

**Papaver rhoeas* L. Corn poppy—Occasional escape from gardens.

**Papaver somniferum* L. Opium poppy—Widespread, common weed of roadsides on San Juan Island; occasional on other islands.

Fumariaceae

**Corydalis lutea* DC. Yellow corydalis—Cultivated and marginally escaping in Friday Harbor and Roche Harbor; an attractive species.

Dicentra formosa (Andr.) Walp. Pacific bleedingheart—Rare; stream valleys on Mt. Constitution; reported also from San Juan Island.

**Fumaria officinalis* L. Common fumitory—Cultivated and occasionally escaping, as on San Juan Island.

Cruciferae

**Alyssum saxatile* L. Yellow alyssum—Garden escape; collected by C.L. Hitchcock, apparently in Friday Harbor. (FH)

**Arabidopsis thaliana* (L.) Schur. Wallcress—An uncommon but widespread weed; European.

**Arabis glabra* (L.) Bernh. Tower mustard—Fairly common weed throughout.

Arabis hirsuta (L.) Scop. Hairy rockcress—Occurs regularly on rocky outcrops and talus slopes. Var. *eschscholtziana* is distributed throughout the lowlands, while var. *glabrata* occurs locally at the summit of Mt. Constitution.

Barbarea orthoceras Ledeb. American wintercress—Low wet areas, often in the vicinity of salt marshes; fairly common.

**Brassica campestris* L. Field mustard—Common; farm fields and disturbed sites. European.

**Brassica juncea* (L.) Coss. Chinese mustard—Occasional weed on larger islands.

**Brassica kaber* (DC.) Wheeler charlock—Common adventive; San Juan and Lopez islands. Less common on other islands.

**Brassica oleracea* L. Cabbage—Occasional garden escape, as on San Juan and Stuart islands.

Cakile edentula (Bigel.) Hook. American searocket—Uncommon on strand; mostly San Juan and Lopez islands.

**Capsella bursa-pastoris* (L.) Medic. Shepherd's-purse—Abundant weed; European.

Cardamine occidentalis (Wats.) Howell Western bittercress—Low moist ground; rare on San Juan Island.

Cardamine oligosperma Nutt. var. *oligosperma* Little Western bittercress—Common native weed; open woods and fields, often where moist.

Cardamine pensylvanica Muhl. Pennsylvania bittercress—Occasional; Cold Springs and Cascade Lake, Orcas Island; rarely found elsewhere; wet areas.

**Cardaria chalapensis* (L.) Hand. Hoarycress—Localized at Mackaye Harbor, Lopez Island, on strand.

Conringia orientalis (L.) Dumort. Hare's-ear—Occasional weed of disturbed wet sites.

**Coronopus didymus* (L.) J.E. Smith Wartcress—European adventive of vacant lots, gardens, etc.; occasional.

Draba verna L. Spring whitlow-grass—Abundant throughout; gravelly meadows, disturbed sites, etc. Two variations are with us; var. *boerhaavii* is abundant and var. *verna* is occasional.

**Hesperis matronalis* L. Damask violet—Fairly common along roadsides on San Juan Island; occasional elsewhere.

Hutchinsia procumbens (L.) Desv. Hutchinsia—Rare; seeps on rocky shorelines, Lopez and San Juan islands. (WTU)

**Lepidium campestre* (L.) R. Br. Field pepperweed—Fields; occasional. European.

Lepidium densiflorum Schrad. Prairie peppergrass—Occasional native weed; disturbed sites, usually where sandy.

Lepidium oxycarpum T.&.G. Sharp-fruited peppergrass—Rare; collected by Dr. Adolf Ceska near Cattle Point in May, 1983 and since relocated; prefers vernal pools; this is the first Washington record for this California taxon. (V) (SA)

**Lepidium perfoliatum* L. Perfoliate peppergrass—Occasional European weed about dumpsters, parking lots, etc.

Lepidium virginicum L. var. *menziesii* Virginia's peppergrass—Locally common along strand and rocky coastal bluffs.

**Lobularia maritima* (L.) Desv. Sweet alison—Escaping garden species in Friday Harbor and Eastsound; European.

**Lunaria annua* L. Money-plant—Occasional escape; mostly San Juan Island.

**Raphanus raphanistrum* L. Jointed charlock—Occasional; European; San Juan and Lopez islands.

**Raphanus sativus* L. Wild radish—Cultivated and marginally spreading at times from gardens, as on San Juan Island.

Rorippa curvisiliqua (Hook.) Bessey var. *lyrata* Western yellowcress—Fairly common; shady wet places, lake margins, often in sandy soils.

Rorippa islandica (Oed.) Borbás var. *islandica* Marshy yellowcress—Rare; wet lake margins, Egg Lake, San Juan Island.

**Rorippa nasturtium-aquaticum* (L.) Schinz. & Thell. Watercress—Low wet ground; scarce.

**Sisymbrium altissimum* L. Tumble mustard—Fairly common European adventive of roadsides and meadows.

**Sisymbrium officinale* (L.) Scop. Hedge mustard—Common weed; disturbed areas. European.

**Teesdalia nudicaulis* (L.) R. Br. Shepherd's cress—Locally common, open lichen balds on Mt. Constitution. (WTU)

**Thlaspi arvense* L. Field pennycress—Occasional European weed.

Thysanocarpus curvipes Hook. Sand fringepod—Rare with us; open, dry gravel slopes. Other than in the Olympic Mountains rainshadow, this species is found almost entirely east of the Cascades. (FS)

Droseraceae

Drosera rotundifolia L. Broad-leaved sundew—Locally abundant, sphagnum bogs on San Juan and Orcas islands; apparently rare on Blakely Island.

Crassulaceae

**Sedum acre* L. Wall-pepper—Widely cultivated and often escaping, particularly from rock gardens in Friday Harbor and East Sound.

**Sedum album* L. White stonecrop—European; widely cultivated and occasionally spreading/escaping.

Sedum lanceolatum Torr. Lance-leaved stonecrop—Widespread, rocky outcrops, usually near sea level, throughout. Var. *lanceolatum* appears to be slightly more common than var. *nesioticum*, a regional endemic.

Sedum oreganum Nutt. Oregon stonecrop—Locally common, open rocky slopes of Mt. Constitution, generally above 1500 feet.

Sedum spathulifolium Hook. Broad-leaved stonecrop—Rocky outcrops, shaded to open; common to abundant.

Tillaea erecta Hook. & Arn. Sand pigmy-weed—Rare; collected by Dr. A. Ceska and H. Janzsen on the west side of San Juan Island; also reported by Dr. Ceska for Speiden Island. These are the only known Washington locations for this California species. (V) (HJ)

Saxifragaceae

Heuchera micrantha Dougl. var. *diversifolia* Small-flowered alumroot—Open to shaded rocky outcrops, especially where steep; common to abundant.

Lithophragma parviflora (Hook.) Nutt. Small-flowered prairie star—Moist meadows; fairly common.

Saxifraga bronchialis L. var. *austromontana* Spotted saxifrage—Locally common; open, windswept rocky slopes on Mt. Constitution and on other higher Orcas Island peaks; less common on San Juan Island and elsewhere.

Saxifraga caespitosa L. var. *subgemmifera* Tufted saxifrage—Widespread on steep, shaded rocky outcrops throughout; most common on Mt. Constitution.

Saxifraga integrifolia Hook. var. *integrifolia* Prairie saxifrage—Moist, spring-laden early spring meadows; common.

Saxifraga occidentalis Wats. var. *rufidula* Western saxifrage—Local; steep, shaded rocky slopes near the summit of Mt. Constitution; scarce elsewhere.

Tellima grandiflora (Pursh) Dougl. Fringe cup—Moist mixed woodlands and shaded valleys; common to abundant.

Tiarella trifoliata L. Western foamflower—Mossy, shaded coniferous woods, and often deciduous or mixed forests as well. Var. *trifoliata* is our common form; var. *laciniata* prefers the most humid, shaded woodlands; less common, mostly on Orcas and Blakely islands.

Grossulariaceae

Ribes divaricatum Dougl. Coast black gooseberry—Forest margins and open transitional woods, often where rocky; fairly common throughout.

Ribes lacustre (Pers.) Poir. Swamp gooseberry—Low, poorly drained situations, often along lake shores or in muddy places; common.

Ribes sanguineum Pursh Red-flowering currant—Dry transitional forests and margins; common.

Hydrangeaceae

Philadelphus lewisii Pursh Mockorange—Steep, rocky slopes and valleys; widespread and locally common.

Rosaceae

Alchemilla occidentalis Nutt. Western lady's mantle—Meadows, gravelly roadsides, and vernal pools; common but inconspicuous.

Amelanchier alnifolia Nutt. var. *semiintegrifolia* Western serviceberry—Open transitional to dry coniferous woodlands, margins, and rocky slopes; fairly common.

Crataegus douglasii Lindl. Black hawthorn—Open country to forest margins or shady woods; widespread but sporadic. Var. *suksdorfii* is most common; var. *douglasii,* more common east of the Cascades, is occasional on dry forest margins. (WWB)

**Crataegus monogyna* Jacq. One-seeded hawthorn—Cultivated and widely escaping; disseminated by birds.

Fragaria chiloensis (L.) Duchesne Coastal strawberry—Strand; occasional, mostly along the southern flank of the archipelago; mostly an outer coast species. (WTU)

Fragaria vesca L. Woodland strawberry—Disturbed, open forests to damp, shady valleys; fairly common to locally abundant. Var. *crinita* is our common form, var. *bracteata* fairly common also but on dryer sites.

Fragaria virginiana Duchesne var. *platypetala* Broad-petaled strawberry—Fairly common in open, moist places to moist woods throughout.

Geum macrophyllum Willd. var. *macrophyllum* Large-leaved geum—Typical of moist, mixed woodlands and stream valleys; tolerant of disturbance. Common.

Holodiscus discolor (Pursh) Maxim. Ocean spray—Abundant; margins and woodlands.

Osmaronia cerasiformis (H. & A.) Indian-plum—Damp alder or mixed woods; scarce, on San Juan Island chiefly, where it is most common in the Cady Mountain and Eagle Cove areas. (FS)

Physocarpus capitatus (Pursh) Kuntze Pacific ninebark—Rare; localized at the n. end of Cascade Lake, Orcas Island, where it grows near the shore. (WTU)

**Potentilla argentea* L. Silvery cinquefoil—European adventive; rare; roadsides on Mt. Dallas. Apparently the first western Washington record (and one of few for the whole state) for this species.

Potentilla gracilis Dougl. var. *gracilis* Slender cinquefoil—Locally common, pastures and roadsides, Friday Harbor and vicinity. (FH)

Potentilla pacifica Howell Pacific cinquefoil—Fresh or salt marshes and coastal beaches; locally abundant.

Potentilla palustris (L.) Scop. Marsh cinquefoil—Riparian to aquatic in shallow ponds and lakes; fairly common.

**Prunus avium* L. Mazzard cherry—Widely cultivated and a well-established, locally common escape in several areas.

Prunus emarginata (Dougl.) Walp. var. *mollis* Bitter cherry—Forests and margins; common.

Prunus virginiana L. var. *demissa* Common chokecherry—Dry coniferous woodlands and edge; fairly common.

Pyrus fusca Raf. Pacific crabapple—Lake edges, rocky slopes, and moist woodlands; widespread and highly variable in habitat.

**Pyrus malus* L. Cultivated apple—Many forms are cultivated and often spread or escape, especially on larger islands.

**Rosa eglanteria* L. Sweetbrier—European; cultivated and sometimes escaping along roadsides, as on Shaw and Orcas islands.

Rosa gymnocarpa Nutt. Little wild rose—Shady woodlands; common.

Rosa nutkana Presl. var. *nutkana* Nootka rose—Sunny, open sites; perhaps the most abundant San Juan shrub.

Rosa pisocarpa Gray Clustered wild rose—Moist meadows and lake or pond edges; occasional. (WS)

**Rubus discolor* Weihe & Nees Himalayan blackberry—Common Eurasian species of roadsides, waste places, etc.

**Rubus laciniatus* Willd. Evergreen blackberry—European; a widespread pest, although less common than *R. discolor.*

Rubus leucodermis Dougl. Blackcap—Open, disturbed woodlands and sunny slopes, especially where formerly grazed; common.

**Rubus macrophyllus* Weihe & Nees Large-leaved blackberry—European; waste places; reported from Yellow Island by The Nature Conservancy.

Rubus parviflorus Nutt. Thimbleberry—Damp, shady woods, often where soils are fertile; common. Occasionally on more exposed slopes and margins as well.

Rubus spectabilis Pursh Salmonberry—Poorly drained woodlands, lake edges, and shady valley bottoms; common.

Rubus ursinus Cham. & Schlecht. var. *macropetalus* Trailing blackberry—Forest margins and disturbed woodlands; abundant.

**Sanguisorba minor* Scop. Garden burnet—European, rarely escaping, as on Orcas Island. (SA)

**Sorbus aucuparia* L. Rowan—Cultivated and often escaping; European.

Spiraea douglasii Hook. Hardhack—Marshy places and other open riparian situations; abundant. Var. *douglasii* is our common form.

Leguminosae

**Coronilla varia* L. Crown vetch—Occasional European weed.

**Cytisus scoparius* (L.) Link Scotch broom—Roadsides and fields throughout; common.

**Laburnum anagyroides* L. Goldenchain—Occasional escape; Friday Harbor and Eastsound.

Lathyrus japonicus Willd. Japanese beach pea—Common on strand.

**Lathyrus latifolius* L. Everlasting pea—An old-world species, occasionally found along roadsides and vacant lots.

Lathyrus littoralis (Nutt.) Endl. Beach pea—Rare; collected by H. Janzsen (1983) on Waldron Island; strand and salt marshes. (HJ)

Lathyrus nevadensis Wats. ssp. *lanceolatus* var. *pilosellus* Sierra Nevada pea—A common forb of dry coniferous forests throughout.

Lathyrus palustris L. Marsh pea—Salt or brackish marshes and strand; widespread but scattered.

**Lotus corniculatus* L. Birdsfoot-trefoil—Common European weed; typically found in low wet places, soggy farm fields, roadsides, etc.

Lotus denticulatus (Drew) Greene Meadow lotus—Steep sandy bluffs along salt water and open meadows; fairly common.

Lotus micranthus Benth. Small-flowered lotus—Roadsides to open woods; often weedy; common.

Lotus purshiana (Benth.) Clements Spanish-clover—Rare; known from a single collection by P. Rapp (1981) from Mt. Constitution. Open prairies; scarce west of the Cascades. (WTU)

Lupinus albicaulis Dougl. Sickle-keel lupine—Localized at Argyle, San Juan Island, where it grows on gravelly glacial outwash deposits; infrequent elsewhere.

**Lupinus arboreus* Sims. Tree lupine—Occasional; introduced and spreading on disturbed, open sites near salt water, principally on Lopez and San Juan islands; native of California.

Lupinus bicolor Lindl. Bicolored lupine—Gravelly prairies throughout; common.

Lupinus latifolius Agardh. var. *latifolius* Broad-leaved lupine—Indigenous populations of this species appear to be restricted to pine-fir woodlands on Mt. Constitution and nearby. Plants found along roadsides elsewhere in the islands are apparently escapees from cultivation.

Lupinus littoralis Dougl. Seashore lupine—Locally abundant on sand dunes at Cattle Point; occasional elsewhere, as on Lopez Island.

Lupinus micranthus Dougl. Small-flowered lupine—Dry, grassy meadows to roadsides; fairly common.

Lupinus microcarpus Sims. var. *scopulorum* Chick lupine—Endemic subspecies; northern Puget Sound. Locally abundant with us; south-facing glacial outwash prairies.

**Medicago lupulina* L. Black medic—Common introduced weed; European.

**Medicago sativa* L. Lucerne—Widespread old-world species of ruderal sites; common and widely introduced.

**Melilotus alba* Desr. White sweet-clover—Common weed; vacant lots and roadsides; European.

**Melilotus officinalis* (L.) Lam. Yellow sweet-clover—Occasional European adventive.

Oxytropis campestris (L.) DC. var. *gracilis* Slender crazyweed—Rare and local; rocky benchland along Iceberg Point, where it can be found occurring with *Armeria maritima* and *Distichlis spicata.* This upper montane species has also been collected on Trial Island south of Victoria. (FS)

Psoralea physodes Dougl. California tea scurf-pea—Rare; collected by Lyall on San Juan Island in 1858; also reported by Baker for several San Juan Island locations. This species is near its northern maximum range in the San Juans.

**Robinia pseudo-acacia* L. Black locust—Escaping from cultivation on occasion, as in Friday Harbor.

**Trifolium cyathiferum* Lindl. Cup clover—Occasional weed of sandy soils; San Juan Island. (WWB)

**Trifolium depauperatum* Desv. Poverty clover—Grassy meadows and bluffs; uncommon.

**Trifolium dubium* Sibth. Least hop clover—Abundant weed along roadsides, fields, and waste areas; European.

**Trifolium hybridum* L. Alsike clover—Common introduced species of lawns and roadsides.

Trifolium macraei H. & A. var. *dichotomum* Macrae's clover—Uncommon; sandy pockets and grassy meadows. Most often collected in the northwestern portion of the archipelago.

Trifolium microcephalum Pursh Woolly clover—Typical of the dry grassy prairies; common throughout.

Trifolium microdon H. & A. Valparaiso clover—Regularly occurring across gravelly sites and open grassy meadows; common throughout.

Trifolium oliganthum Steud. Few-flowered clover—Locally common on small pockets of soil in rocky areas, in addition to being a regular element of Red fescue prairies.

**Trifolium pratense* L. Red clover—One of our most common clovers, found throughout the islands in disturbed places.

**Trifolium procumbens* L. Hop clover—Native of Europe; common throughout in waste places.

**Trifolium repens* L. White clover—Well-established and abundant throughout; ruderal sites.

**Trifolium subterraneum* L. Subterranean clover—Intruced pasture species; widespread but occasional in well-grazed farm fields and along roadsides.

Trifolium tridentatum Lindl. Tomcat clover—Our most abundant native clover; dry grassy prairies and rocky benchland throughout.

Trifolium variegatum Nutt. White-tip clover—Fairly common; generally occupies the mesic and somewhat disturbed portions of grass prairies.

Trifolium wormskjoldii Lehm. Springbank clover—Rare; two old collections from San Juan and Stuart islands and one from the summit of Mt. Constitution. Generally an outer coast species in Washington; prefers strand and adjacent meadows. (WTU)

**Ulex europaeus* L. Gorse—Introduced and occasionally escaping, as on Orcas Island. A terrible pest from Europe, now common in much of Washington.

Vicia americana Muhl. American vetch—A very common species of forest margins and thickets. Material in the San Juans is variable, representing two forms, vars. *truncata* and *villosa.*

Vicia gigantea Hook. Giant vetch—Common next to salt water in thickets and open forests.

**Vicia hirsuta* (L.) S.F. Gray Hairy vetch—Abundant European weed of roadsides and fields.

**Vicia lathyroides* L. Pea vetch—Occasional European species, chiefly on strand.

**Vicia sativa* L. Common vetch—Very common throughout; waste areas and cultivated ground.

**Vicia tetrasperma* L. Slender vetch—Occasional; weedy sites, especially on San Juan Island.

**Vicia villosa* Roth Woolly vetch—Common; disturbed sites.

Geraniaceae

**Erodium cicutarium* (L.) L'Her. Crane's bill—Gravelly meadows and rocky outcrops; common.

Geranium bicknellii Britt. Bicknell's geranium—Moist forest margins and damp meadows; occasional.

**Geranium carolinianum* L. Carolina geranium—Fairly common in waste places.

**Geranium dissectum* L. Cut-leaved geranium—European weed; widespread but most common on San Juan Island.

**Geranium molle* L. Dovefoot geranium—European; common weed throughout.

**Geranium pusillum* Burm. Small-flowered geranium—European adventive; fairly common on lawns, in gardens, etc.

**Geranium robertianum* L. Robert geranium—Introduced Eurasian species, occasionally escaping, as on Waldron Island.

Oxalidaceae

**Oxalis corniculata* L. Creeping yellow wood-sorrel—Eurasian adventive; occasionally present in gardens, as in Friday Harbor and East Sound.

Euphorbiaceae

**Euphorbia cyparissias* L. Cypress spurge—Eurasian species sometimes escaping from rock gardens; present chiefly in Friday Harbor.

**Euphorbia peplus* L. Petty spurge—Occasional European weed.

Callitrichaceae

Callitriche anceps Fern. Two-edged water-starwort—Aquatic of shallow, muddy-bottomed ponds; San Juan Island; rare.

Callitriche heterophylla Pursh var. *bolanderi* Varied-leaved water-starwort—Aquatic of shallow ponds and lakes; most common on Orcas Island but also on San Juan Island.

Callitriche verna L. Spring water-starwort—Occasional aquatic, larger lakes. (WTU)

Thymelaceae

**Daphne laureola* L. Spurge-laurel—An introduced species, occasionally escaping and establishing itself in shady woods.

Anacardiaceae

**Rhus glabra* L. Smooth sumac—An occasional escape of roadsides in Friday Harbor. (SA)

Celastraceae

Pachistima myrsinites (Pursh) Raf. Mountain-lover—Steep, shaded rocky slopes; mostly confined to Orcas and the Sucia Island group but occasionally found elsewhere.

Aceraceae

Acer glabrum Torr. var. *douglasii* Douglas' maple—Moist valleys and open rocky slopes, widespread and fairly common throughout.

Acer macrophyllum Pursh Bigleaf maple—Dry to moist slopes; common.

Rhamnaceae

Rhamnus purshiana DC. Cascara—An understory species of dry to moist coniferous forest and margins; rather rare and scattered through the archipelago on the larger island. (SA)

Malvaceae

**Althea rosea* L. Hollyhock—Escaping from its garden borders, principally on Orcas Island; European.

**Malva moschata* L. Musk mallow—Garden escape, European; occasional on main islands.

**Malva neglecta* Wallr. Dwarf mallow—Occasional European adventive.

**Malva sylvestris* L. Common mallow—Fairly common European adventive of planted gardens, hard-packed waste soils, etc., on the larger islands.

Sidalcea hendersonii Wats. Henderson's sidalcea—Grassy salt marshes; Garrison-Wescott Bays near Roche Harbor, San Juan Island, and on Henry Island. (WTU)

Hypericaceae

Hypericum anagalloides C. & S. Bog St. John's wort—Swampy lake shores and bogs; fairly common on Orcas and San Juan islands.

Hypericum formosum H.B.K. var. *scouleri* Western St. John's wort—Bogs to soggy ditches; uncommon.

**Hypericum perforatum* L. Common St. John's wort—Common Eurasian weed.

Violaceae

Viola adunca Sm. var. *adunca* Western long-spurred violet—Widespread on meadows; locally abundant near Cattle Point, San Juan Island.

Viola glabella Nutt. Stream violet—Rare and local; moist meadows on Mt. Constitution and, at least formerly, on Lopez Island. (FS)

Viola howellii Gray Howell's violet—Dry coniferous forest borders and scrubby adjacent meadows; occasional. (SA)

Viola mackloskeyi Lloyd var. *mackloskeyi* Mackloskey's violet—Local; Killebrew Lake bog. (FS)

Viola palustris L. Marsh violet—Locally common, streamsides on Mt. Constitution, generally above 1,000 feet.

Viola sempervirens Greene Evergreen violet—Locally common, mossy coniferous forest floors on Mt. Constitution and nearby; scarce elsewhere.

Cactaceae

Opuntia fragilis (Nutt.) Haw. Prickly pear cactus—Widespread on dry rocky outcrops or meadows, most often near salt water; particularly numerous on Lopez and several adjacent small islands. (WTU)

Elaeagnaceae

Shepherdia canadensis (L.) Nutt. Soopolallie—Dry coniferous and open transitional woodlands; particularly common on Lopez Island but widespread throughout.

Lythraceae

**Lythrum salicaria* L. Purple loosestrife—Introduced around ponds and in gardens on San Juan Island, where also infrequently escaping; becoming a nuisance in other parts of western Washington, so discretion should be used in planting it.

Onagraceae

Boisduvalia densiflora (Lindl.) Wats. Dense spike-primrose—Rare; open, wet places; collected by Peck near Friday Harbor in 1923. (WS)

Circaea alpina L. Enchanter's nightshade—Common; cool, shady woods.

Clarkia amoena (Lehm.) Nels. & Macbr. Farewell-to-spring—Open, less disturbed meadows; widespread but most common in the northern part of the archipelago, from Stuart and adjacent small islets east to Sucia and Orcas islands. Var. *caurina* is our common form, var. *lindleyi* occasional, and sometimes found with var. *caurina,* as on Stuart Island and Mt. Constitution. Highly variable in form and a confusing species. A Peck collection mis-identified as *C. gracilis* (Piper) Nels. & Macbr. at WTU is referable to *C. caurina* var. *lindleyi.*

Epilobium alpinum L. var. *alpinum* Alpine willow-herb—Moist meadows and seeps on Mt. Constitution; rare. (WTU) (WWB)

Epilobium angustifolium L. Fireweed—Forest margins, roadsides, and waste places; common to abundant.

Epilobium minutum Lindl. Small-flowered willow-herb—Open rocky outcrops; fairly common.

Epilobium palustre L. Swamp willow-herb—Sphagnum bogs on Orcas Island; Killebrew and Summit lakes; rare.

Epilobium paniculatum Nutt. var. *paniculatum* Autumn willow-herb—Dry, hard-packed soils and waste places; fairly common.

Epilobium watsonii Barbey Watson's willow-herb—Var. *watsonii* is common throughout in poorly drained places, ditches, etc., often where it is weedy. Var. *parishii* is less common in similar habitats.

Ludwigia palustris (L.) Ell. Water-purslane—Shallow water of Cascade and Killebrew lakes, Orcas Island, and on Blakely Island at Spencer Lake; probably will be found elsewhere.

Oenothera contorta (Dougl.) Kearney Contorted-pod evening-primrose—Sand dunes and rocky bluffs; fairly common at S. Beach and adjacent Cattle Point, San Juan Island; scarce from American Camp to Mt. Dallas. Apparently restricted to the San Juan islands and at a few places along the Columbia River in Washington. (WTU)

**Oenothera erythrosepala* Borb. Red-sepaled evening-primrose—An ornamental hybrid species, well-established and locally abundant at Argyle, south of Friday Harbor; scarce elsewhere.

Haloragaceae

**Myriophyllum spicatum* L. var. *exalbescens* Eurasian water-milfoil—Common aquatic; becoming a nuisance in some areas.

Hippuridaceae

Hippuris montana Ledeb. Mt. mare's tail—Sphagnum mats; rare; chiefly Summit Lake on Mt. Constitution.

Hippuris vulgaris L. Common mare's tail—Common aquatic, sometimes overrunning shallow shorelines.

Araliaceae

**Hedera helix* L. English ivy—Often planted and frequently spreading.

**Oplopanax horridum* (Smith) Miq. Devil's club—Forestry introduction in stream valleys near Cascade Lake, Orcas Island; according to Baker and others, plants were established there in 1947; no longer present.

Umbelliferae

**Anthriscus scandicina* (Weber) Mansfield Bur chervil—Lawns and roadsides; occasional European adventive.

**Carum carvi* L. Caraway—Cultivated and occasionally turning up along roadsides, as at Roche Harbor and near Argyle, San Juan Island.

Caucalis microcarpa H. & A. California hedge-parsley—Damp meadows and vernal pools; rare; San Juan Island. (WTU) (WWB)

Cicuta douglasii (DC.) Coult. & Rose Douglas water-hemlock—Locally common riparian and shallow water species, Sportsmans Lake, San Juan Island. Deadly poisonous. (FH)

Conioselinum pacificum (Wats.) Coult. & Rose Pacific hemlock-parsley—Maritime environments, particularly shady coves and cobble or glacial outwash beaches; fairly common throughout.

**Conium maculatum* L. Poison-hemlock—European adventive of vacant lots and roadsides; common and apparently increasing. Deadly poisonous.

**Daucus carota* L. Queen Anne's lace—Perhaps the most abundant European adventive in the San Juans.

Daucus pusillus Michx. Rattlesnake weed—Rocky outcrops; fairly common but erratic.

**Foeniculum vulgare* Mill. Sweet fennel—Well-established Mediterranean weed near Argyle Lagoon, San Juan Island, and occasionally elsewhere.

Glehnia leiocarpa Mathias. American glehnia—Strand; rare; mostly on southerly beaches of San Juan and Lopez islands; not always present, and only represented by scattered specimens some years.

Heracleum lanatum Michx. Cow-parsnip—Coastal bluffs, wet, open valleys, and maritime environments; common, especially on smaller islands.

**Heracleum mantegazzianum* Somm. & Lev. Giant hog fennel—Occasional escapee from cultivation; Orcas Island.

Lilaeopsis occidentalis Coult. & Rose Lilaeopsis—Muddy, rocky pools just above the high tide mark; rare; Eagle Cove, San Juan Island. (FH)

Lomatium nudicaule (Pursh) Coult. & Rose Naked desert-parsley—Moist meadows to grassy beaches; common.

Lomatium utriculatum (Nutt.) Coult. & Rose Spring gold—Moist meadows; locally abundant and widespread.

Oenanthe sarmentosa Presl. Water-parsley—Damp, low woodland sites, stream valleys, sedge fens, and ditches; common.

Osmorhiza chilensis H. & A. Mt. sweet-cicely—Dry to moist shady coniferous forests; common to abundant.

Osmorhiza purpurea (Coult. & Rose) Suksd. Purple sweet-cicely—Dense, moist coniferous woods; Mt. Constitution and adjacent lowland areas in Moran State Park, such as Mountain Lake; also on Blakely Island; occasional. Very similar to *O. chilensis*.

**Pastinaca sativa* L. Common parsnip—European garden species, infrequently escaping/spreading, as in Friday Harbor.

Perideridia gairdneri (H. & A.) Math. Gairdner's yampah—Locally abundant, less disturbed prairies; particularly numerous at Iceberg Point, Lopez Island.

**Petroselinum crispum* (Mill.) A.W. Hill Cultivated parsley—European cultivar and occasionally escaping, as in Friday Harbor and at Orcas, Orcas Island.

**Pimpinella saxifraga* L. Burnet-saxifrage—Eurasian weed; collected by Morton E. Peck near Roche Harbor in 1923. This is apparently the only Washington record for this species.

Sanicula arctopoides H. & A. Footsteps-of-spring—Coastal bluffs and consolidated dunes; rare, collected by Lyall on Orcas Island in 1858 (Piper, 1906). In its Washington range, now only present at Leadbetter Point, Pacific County.

Sanicula bipinnatifida Dougl. Purple sanicle—Dry, glacial outwash meadows; occasional, although locally common at Kanaka Bay, San Juan Island.

Sanicula crassicaulis Poepp. var. *crassicaulis* Pacific sanicle—Dry transitional coniferous woods; common.

Cornaceae

Cornus nuttallii Aud. Pacific dogwood—Scattered and rare on larger islands in mixed alder-fir forests; more common on Stuart, Waldron, and other northern islands. (FH)

Cornus stolonifera Michx. var. *occidentalis* Red-osier dogwood—Swampy places in Humic Gleyed soils; widespread and fairly common near lakes.

Cornus unalaschkensis (Ledeb.) Bain & Denf. *(C. canadensis* L.) Bunchberry—Moist, mossy coniferous woods near Summit Lake, Mt. Constitution; rare. (WTU) (FH)

Ericaceae

Allotropa virgata T. & G. Candystick—Rare; deep coniferous woods; reported from San Juan and Orcas islands.

Arbutus menziesii Pursh Pacific madrona—Common; coastal bluffs and rocky upland areas.

Arctostaphylos columbiana Piper Hairy manzanita—Locally common; rocky outcrops in pine forests near the summit of Mt. Constitution. (WTU)

Arctostaphylos uva-ursi (L.) Spreng. Kinnikinnick—Cultivated and occasionally escaping; apparently rare as a native, as on Stuart Island; rocky, exposed slopes.

Chimaphila menziesii (R. Br.) Spreng. Little prince's-pine—Uncommon and local in moist woods; Mt. Constitution; scarce elsewhere.

Chimaphila umbellata (L.) Bart. var. *occidentalis* Pipsissewa—Locally common, moist pine and fir woods, Mt. Constitution. Rare on San Juan Island and elsewhere.

Gaultheria shallon Pursh Salal—Abundant; dry to moist woods throughout.

Hypopitys monotropa Crantz. Fringed pinesap—Widespread in deep coniferous woods; most common on Mt. Constitution.

Kalmia occidentalis Small Western swamp-laurel—Localized; sphagnum bogs on San Juan and Orcas islands.

Ledum groenlandicum Oeder Bog Labrador-tea—Locally common; bogs on San Juan and Orcas islands.

Monotropa uniflora L. Indian-pipe—Fairly common; deep coniferous woods throughout.

Pterospora andromedea Nutt. Pinedrops—Old-growth coniferous forests to open fir woods; occasional. (FH) (WTU)

Pyrola aphylla Smith Leafless pyrola—Occasional; deep coniferous woods throughout, but most common on Orcas Island.

Pyrola asarifolia Michx. Pink wintergreen—Deep coniferous woods; uncommon and local on Mt. Constitution. Rare elsewhere. (WTU)

Pyrola chlorantha Sw. Green wintergreen—Deep, damp woods on Mt. Constitution; uncommon. Locally more common in the vicinity of Cascade and Mountain Lakes; rare on San Juan Island. (WWB)

Pyrola dentata Smith Dentate pyrola—Rare; deep coniferous woods; collected by A. R. Roos on Mt. Constitution in 1910. (WWB)

Pyrola minor L. Lesser wintergreen—Occasional; deep coniferous forest floors to moist, gravelly stream valleys on Mt. Constitution.

Pyrola secunda L. var. *secunda* One-sided wintergreen—Uncommon and scattered; boggy lake margins and mossy coniferous woods on Mt. Constitution; rare on San Juan Island.

Pyrola uniflora L. Wood-nymph—Reported by Walter Harm from shady, damp woods between Mountain and Twin lakes in Moran State Park, Orcas Island.

Vaccinium ovatum Pursh Evergreen huckleberry—Rare and local in the northwest portion of the archipelago; dry, open fir woodlands. (FS)

Vaccinium parvifolium Smith Red huckleberry—Uncommon in moist woods, often growing from decaying stumps; Blakely and Orcas islands, and becoming more common as an understory component in the Sucia Island group.

Vaccinium uliginosum L. Bog bilberry—Rare; Summit Lake on Orcas Island; sphagnum islets. Originally collected by Henderson in 1892. (WTU)

Primulaceae

**Anagallis arvensis* L. Poor man's weatherglass—Locally common European weed along roadsides, San Juan Island.

Dodecatheon hendersonii Gray Henderson's shooting star—Locally common, Mt. Dallas region, San Juan Island; occasional elsewhere. (WTU)

Dodecatheon pulchellum (Raf.) Merrill var. *pulchellum* Few-flowered shooting star—Common; grassy meadows where moist in spring; often on bluffs above the salt water and extending up to higher elevations on Mt. Constitution.

Glaux maritima L. Sea milkwort—Occasional in salt marshes, Lopez Island. (WWB)

Lysimachia thyrsiflora L. Tufted loosestrife—Occasional; shallow water of lakes and ponds, chiefly Orcas Island. (WTU)

Trientalis arctica Fisch. Northern starflower—Locally abundant, Beaverton Marsh, San Juan Island; also common in sphagnum bogs on Orcas Island; scarce elsewhere.

Trientalis latifolia Hook. Broad-leaved starflower—Abundant throughout; shady woods.

Plumbaginaceae

Armeria maritima (Mill.) Willd. Sea-pink—Locally abundant on open rocky and gravelly sites, typically along salt water.

Buddlejaceae

**Buddleja davidii* Franch. Butterfly-bush—Introduced and occasionally escaping, as at Lopez Village.

Gentianaceae

**Centaurium erythraea* Gilib. Common centaury—Locally abundant weed; San Juan Island. Less common on other major islands. Listed in Hitchcock as *C. umbellatum.*

Gentiana amarella L. Northern gentian—Moist meadows, often where disturbed; widespread but scattered; all major islands.

Menyanthaceae

Menyanthes trifoliata L. Buckbean—Shallow, muddy lake edges; most common on Orcas but also on San Juan Island.

Apocynaceae

Apocynum androsaemifolium L. var. *pumilum* Spreading dogbane—Localized on dry, open rocky slopes and open woodlands on Mt. Constitution; isolated near False Bay, San Juan Island.

**Vinca major* L. Periwinkle—Cultivated and marginally escaping/spreading, as in East Sound.

Convolvulaceae

**Convolvulus arvensis* L. Field morning glory—European weed, locally common on San Juan Island and less common on other larger islands.

**Convolvulus sepium* L. Hedge morning glory—European weed, common in waste places, particularly in towns.

Convolvulus soldanella L. Beach morning glory—Locally common, strand and glacial outwash bluffs from Cattle Point to South Beach. An outer coast species, this morning glory is mostly absent from Puget Sound.

Cuscutaceae

Cuscuta epithymum Murr. Common dodder—Rare; parasitic on legumes in our area; collected by Peck near Roche Harbor in 1923. (WTU)

Cuscuta salina Engelm. Saltmarsh dodder—Salicornia marshes; common.

Polemoniaceae

Collomia grandiflora Dougl. Large-flowered collomia—Dry, gravelly prairies and roadsides; reported by Higgenbotham for the interior of San Juan Island.

Collomia heterophylla Hook. Varied-leaved collomia—Dry open forests and margins; widespread but inconspicuous and scattered.

Gilia capitata var. *capitata* Sims. Globe gilia—Rare; gravelly meadows and roadsides; collected by Peck near Friday Harbor in 1923.

Linanthus bicolor (Nutt.) Greene var. *minimus* Bicolored linanthus—Scrubby meadows to dry, open coniferous woods; occasional throughout. (WTU)

Microsteris gracilis (Hook.) Greene Pink microsteris—Occasional in meadows; Orcas and islands towards the north. Var. *humilior,* more common east of the Cascades, is present at Mt. Constitution, although it is rare there (WTU), var. *gracilis* is more frequent on Orcas and in the Sucia Island group.

Navarretia minima Nutt. Least navarretia—Dry, gravelly prairies; although not reported by Hitchcock to occur west of the Cascades, reported by Higinbotham from the interior of San Juan Island.

Navarretia squarrosa (Esch.) H. & A. Skunkweed—Locally abundant and widespread; gravelly roadsides and meadows, particularly on San Juan Island.

Polemonium pulcherrimum Hook. var. *pulcherrimum* Showy polemonium—Open, steep rocky slopes on Mt. Constitution; rarely found elsewhere (e.g., at Iceberg Point, Lopez Island).

Hydrophyllaceae

Nemophila parviflora Dougl. var. *parviflora* Small-flowered nemophila—Damp, shady woods, usually of alder or of mixed composition; common.

Phacelia linearis (Pursh) Holz. Thread-leaved phacelia—Dry, open rock faces, usually growing in crevices; rare on San Juan Island. Found mainly east of the Cascades.

Boraginaceae

Amsinckia intermedia Fisch. & Mey. Ranchers' tarweed—Fairly common; dry, hard-packed waste soils; chiefly San Juan and Lopez islands. (FH)

Amsinckia menziesii (Lehm.) Nels. & Macbr. Menzies' tarweed—Dry meadows, coastal strand, and waste places; common throughout.

**Anchusa azurea* Mill. Italian bugloss—Very dry glacial outwash meadows and roadsides; rare on San Juan Island.

**Anchusa officinalis* L. Common bugloss—Introduced weed, common in Friday Harbor; occasional elsewhere; European.

**Borago officinalis* L. Common borage—European cultivated species; occasionally escaping from gardens.

**Myosotis arvensis* (L.) Hill Field forget-me-not—Common European weed.

**Myosotis discolor* Pers. Yellow-and-blue forget-me-not—Rocky meadows and disturbed places; abundant but easily overlooked; European.

Myosotis laxa Lehm. Small-flowered forget-me-not—Wet places, mostly on pond and lake shores; common throughout.

**Myosotis micrantha* Pall. Blue forget-me-not—Meadows to gardens and roadsides; uncommon adventive from Europe.

Plagiobothrys scouleri (H. & A.) Johnst. Scouler's popcorn-flower—Vernal pools and moist meadows; widespread and fairly common. Var. *scouleri* is the typical upland form; var. *penicillatus,* the more prostrate form, is occasional along grassy beaches, as at Spencer Spit, Lopez Island. This form is more common east of the Cascades. (WWB)

Plagiobothrys tenellus (Nutt.) Gray Slender popcorn-flower—Dry upland meadows, mostly on San Juan Island, where occasional; rare elsewhere. Found chiefly east of the Cascades. (WTU)

**Symphytum officinale* L. Common comfrey—Cultivated and now widespread as a roadside species on all main islands; European.

Labiatae

**Ajuga reptans* L. Bugle—Commonly planted as an ornamental and infrequently escaping/spreading, as in Friday Harbor and East Sound.

**Glecoma hederacea* L. Creeping charlie—Occasional Eurasian adventive of lawns.

**Lamium amplexicaule* L. Henbit dead-nettle—Occasional, Orcas and Lopez islands.

**Lamium maculatum* L. Spotted dead-nettle—Rare European adventive, thus far collected only near Orcas, Orcas Island, and at Lopez Village. (SA)

**Lamium purpureum* L. Red dead-nettle—Common early spring adventive.

Lycopus uniflorus Michx. Northern water-horehound—Lakeshores, bogs, and on fallen logs near fresh water; widespread and fairly common.

**Marrubium vulgare* L. Horehound—Open meadows with a past history of grazing or other disturbance; sporadic.

**Melissa officinalis* L. Lemon balm—Cultivated and widely naturalized, now fairly common on a number of roadsides; European.

Mentha arvensis L. Field mint—Riparian habitats; common to abundant.

**Mentha citrata* Ehrh. Bergamot mint—Stream valleys to lakeshores; scarce, mostly on San Juan Island.

Mentha piperata L. Peppermint—Lakeshores to roadsides; occasional throughout.

**Mentha rotundifolia* (L.) Huds. Apple mint—Occasional in low, disturbed places; Orcas Island. (FS)

Mentha spicata L. Spearmint—Grassy areas, lakeshores, often weedy; occasional.

**Nepeta cataria* L. Catnip—Eurasian; widespread along roadsides on San Juan Island; less common on other islands.

Prunella vulgaris L. var. *lanceolata* Self-heal—Woodland margins and meadows, often weedy; common to abundant throughout.
Satureja douglasii (Benth.) Briq. Yerba buena—Dry to moist coniferous forests and margins; common.
Scutellaria galericulata L. Marsh skullcap—Lake shorelines; fairly common on Orcas Island; less common on San Juan and Blakely islands.
Stachys cooleyae Heller Cooley's hedge-nettle—Damp thickets and low, shaded swampy ground; common throughout.
Thymus serpyllum L. Garden thyme—Introduced and rarely escaping, as at the Shaw Island ferry landing.

Solanaceae

**Solanum dulcamara* L. Climbing nightshade—Eurasian; occasional on larger islands along lakeshores, waste places, etc.
**Solanum nigrum* L. var. *virginicum* Black nightshade—Occasional weed of gardens and farmland.
**Solanum sarrachoides* Sendt. Hairy nightshade—Native of South America; common weed of gardens, sandy meadows, etc.

Scrophulariaceae

**Antirrhinum majus* L. Common snapdragon—Commonly introduced and often escaping, as on San Juan and Orcas islands.
Castilleja hispida Benth. var. *hispida* Hairy Indian paintbrush—Grassy meadows, steep, moist rocky slopes, and occasionally forest margins; widespread but erratic.
Castilleja levisecta Greenm. Golden Indian paintbrush—Rare; collected in 1917, 1923, and 1936 between Friday Harbor, Cattle Point, and Kanaka Bay on San Juan Island; a small population has been recently found on southern San Juan Island. This is a rare regional endemic, found from southern Vancouver Island to the Willamette Valley, Oregon, west of the Cascades. Only 6 populations are known from Washington. A beautiful species. (WTU)
Collinsia grandiflora Lindl. Large-flowered collinsia—Forest openings to grassy bluffs; occasional.
Collinsia parviflora Lindl. Blue-eyed Mary—Moist meadows and rocky outcrops; abundant.
**Cymbalaria muralis* Gaertn. Kenilworth ivy—Eurasian ornamental; occasional escape from rock gardens.
**Digitalis purpurea* L. Foxglove—Disturbed woodlands and edge; common to abundant; European.
Linaria canadensis (L.) Dumont var. *texana* Blue toadflax—Moist, shady slopes; rare. (WWB)
**Linaria dalmatica* (L.) Mill. Dalmatian toadflax—Occasional European adventive, San Juan Island.
**Linaria vulgaris* Hill Butter and eggs—Occasional garden weed, Friday Harbor.
Mimulus alsinoides Dougl. Chickweed monkey flower—Moist rocky seeps, often where steep; most common on Orcas Island but widespread throughout.
Mimulus guttatus DC. Common monkey flower—Wet, open places and lakeshores, to coastal seeps; common. Var. *guttatus* is our most common form; var. *depauperatus* is widespread but scattered in steep, rocky seeps, often growing with *M. alsinoides*. Var. *grandis* is restricted to wet coastal bluffs, chiefly on San Juan Island.
Mimulus moschatus Dougl. var. *sessilifolius* Coastal musk flower—Damp woods and riparian situations; rare on San Juan Island.
Orthocarpus attenuatus Gray Narrow-leaved owl-clover—Meadows; widespread but scattered. (WTU)
Orthocarpus bracteosus Benth. Rosy owl-clover—Meadows; rare; collected at two locations on San Juan Island. (FH) (WTU)
Orthocarpus pusillus Benth. Dwarf owl-clover—Meadows to weedy pastures and lawns; common.
**Parentucellia viscosa* (L.) Car. Yellow bartsia—Moist ditches and wet, weedy pastures; common to locally abundant; Mediterranean.
**Verbascum blattaria* L. Moth mullein—Occasional Eurasian escape, San Juan Island.
**Verbascum thapsus* L. Flannel mullein—Roadsides and fields; common European adventive.
Veronica americana Schwein. American speedwell—Open, wet places throughout; common.
**Veronica arvensis* L. Wall speedwell—European; abundant but inconspicuous adventive of rocky outcrops, meadows, and roadsides.
**Veronica chamaedrys* L. Germander speedwell—European adventive of lawns; occasional.

Veronica filiformis Sm. Slender speedwell—European adventive of lawns; occasional.
**Veronica officinalis* L. Common speedwell—European; forest margins, roadsides, etc.; common, especially where moist.
Veronica peregrina L. Purslane speedwell—Vernal pools to open marshy places; rather rare. Var. *peregrina*, an introduced species, is local at Kanaka Bay, San Juan Island; var. *xalapensis* is scarce, known from Stuart and Orcas islands only. (WTU)
Veronica scutellata L. Skullcap speedwell—Riparian to semi-aquatic; common.
**Veronica serpyllifolia* L. Thyme-leaved speedwell—Moist meadows, roadsides and waste places; common. Var. *serpyllifolia* is the common European adventive of weedy sites; var. *humifusa* is an occasional native of wet, open places.

Orobanchaceae

Boschniakia hookeri Walpers. Hooker's ground-cone—Parasitic on Salal; rare; reported from Orcas and San Juan islands in deep coniferous woods.
Orobanche californica Cham. & Schlecht. California broomrape—Open rocky slopes, where parasitic on *Grindelia integrifolia* near salt water; fairly common, especially on smaller islets. Var. *californica* is our common form; var. *grayana* is scarce on northern islets. (WWB)
Orobanche uniflora L. var. *minuta* Naked broomrape—Open rocky outcrops, locally common; parasitic on stonecrops.

Lentibulariaceae

Utricularia vulgaris L. Common bladderwort—Common aquatic; deep to shallow water.

Plantaginaceae

Plantago elongata Pursh Slender plantain—Rocky bluffs, strand, and vernal pools; locally abundant, as at Cattle Point. (WTU)
**Plantago lanceolata* L. Buckhorn plantain—Common European weed.
**Plantago major* L. Common plantain—Var. *major* is abundant along roadsides, wet places, ditches, etc.; European. Var. *pachyphylla* is a widespread native of rocky shorelines and sand dunes. (WTU)
Plantago maritima L. ssp. *juncoides* Sea plantain—Rocky coastlines and salt marshes; common throughout.

Rubiaceae

**Asperula odorata* L. Sweet woodruff—European garden species, occasionally escaping, as in Friday Harbor.
**Galium aparine* L. Catchweed bedstraw—Abundant; waste places to open woodlands. Var. *echinospermum* is our common form, var. *aparine* occasionally intermixed with it.
Galium boreale L. Northern bedstraw—Damp, flat meadows; local on southern San Juan Island from Cattle Point to Kanaka Bay, San Juan Island. (WTU)
Galium cymosum Wieg. Pacific bedstraw—Gravelly meadows, moist; occasional. (WTU)
**Galium mollugo* L. Wild madder—Eurasian; occasional weed, San Juan Island. (FH)
Galium trifidum L. var. *pacificum* Small bedstraw—Marshy places and lake edges; abundant.
Galium triflorum Michx. Fragrant bedstraw—Dry to moist coniferous woods; abundant.
**Sherardia arvensis* L. Blue field-madder—Eurasian weed of grazed meadows and lawns; sporadic.

Caprifoliaceae

Linnaea borealis L. var. *longiflora* Twinflower—Moist, shady coniferous woods, often where mossy; common.
Lonicera ciliosa (Pursh) DC. Orange honeysuckle—Open transitional and dry coniferous forests; common.
Lonicera hispidula (Lindl.) Dougl. Hairy honeysuckle—Open transitional and dry coniferous forests; common.
Lonicera involucrata (Rich.) Banks var. *involucrata* Black twinberry—Swampy, low ground, often along lakes; fairly common. Commonly growing with *Cornus stolonifera*.
Sambucus cerulea Raf. Blue elderberry—Rare; open rocky slopes, Mt. Dallas and vicinity.

Sambucus racemosa L. var. *arborescens* Red elderberry—Damp, shady woods to sunny, open slopes; common.
Symphoricarpos albus (L.) Blake var. *laevigatus* Snowberry—Dry forests to roadsides; abundant.
Symphoricarpos mollis Nutt. Creeping snowberry—Moist mixed woodland slopes; rare; chiefly Mt. Dallas-Cady Mountain. (SA)

Valerianaceae

**Centranthus ruber* (L.) DC. Juniper's beard—Garden species; occasionally escaping, as in Friday Harbor.
Plectritis congesta (Lindl.) DC. Sea blush—Moist meadows to grassy beaches; locally abundant.
Valeriana scouleri Rydb. Scouler's valerian—Moist, shady coniferous woods; rare; confined in our area to Sucia Island.
**Valerianella locusta* (L.) Betcke Corn salad—Thinned coniferous forests and margins; occasional but locally abundant at Lopez Hill. European.

Dipsacaceae

**Dipsacus sylvestris* Huds. Teasel—European adventive; fairly common along roadsides.

Cucurbitaceae

Marah oreganus (T. & G.) Howell Oregon manroot—Gravelly meadows, scrubby areas, and rocky slopes; mostly on San Juan and Orcas islands, where locally common.

Campanulaceae

**Campanula medium* L. Canterbury bell—European garden escape; occasional along roadsides in Friday Harbor and East Sound.
**Campanula persicifolia* L. Garden bluebell—European adventive; occasional.
**Campanula rapunculoides* L. Creeping bellflower—Occasional garden escape along cracks in pavement and on roadsides, as in Friday Harbor and elsewhere.
Campanula rotundifolia L. Bluebells-of-Scotland—Steep, shaded rocky slopes; widespread but most common on Mt. Constitution and at Chadwick Hill, Lopez Island.
Campanula scouleri Hook. Scouler's harebell—Moist coniferous forests; widespread and fairly common, especially on Orcas Island.
Triodanis perfoliata (L.) Nieuwl. Venus' looking glass—Steep, exposed rocky slopes; widespread but scattered, principally on Lopez, San Juan, Stuart and islands in the Yellow Island group. (FS)

Compositae

Achillea millefolium L. Yarrow—Abundant; meadows, bluffs, strand, and weedy sites. Our material is variable, conforming to several subspecific taxa not to be described herein.
Adenocaulon bicolor Hook. Pathfinder—Shady woods throughout; abundant.
Agoseris grandiflora (Nutt.) Greene Large-flowered agoseris—Fairly common throughout on prairies; locally common on Mt. Constitution. Highly variable.
Agoseris heterophylla (Nutt.) Greene Annual agoseris—Occasional on dry to moist meadows. (WTU)
Ambrosia chamissonis (Less.) Greene var. *bipinnatisecta* Silver bursage—Common; strand.
Anaphalis margaritacea (L.) B. & H. Pearly everlasting—Abundant; roadsides, strand; rather weedy.
Antennaria microphylla Rydb. Rosy pussytoes—Local; Mt. Constitution; rocky, semi-open places.
Antennaria neglecta Greene Field pussytoes—Occasional; dry forest margins.
**Anthemis arvensis* L. Field chamomile—Occasional weed of towns and strand; Eurasian.
**Anthemis cotula* L. Stinking mayweed—Locally common weed in Friday Harbor; occasional elsewhere; Eurasian.
**Arctium minus* (Hill) Bernh. Burdock—Occasional Eurasian weed.
**Artemisia absinthium* L. Wormwood—Occasional adventive of sandy soils; mostly on San Juan Island.
Artemisia campestris L. ssp. *borealis* var. *scouleriana* Northern wormwood—Locally common

but erratic; highly variable in habitat; glacial outwash meadows, strand, dry rocky slopes, etc.

Artemisia suksdorfii Piper Coastal mugwort—Fairly common; glacial bluffs and beaches.

Artemisia tilesii Ledeb. var. *unalaschensis* Aleutian mugwort—Localized on high rocky ridges at the summit of Mt. Constitution; fairly common. (FS)

Aster chilensis Nees Chilean aster—Common throughout, often where moist. Highly variable in form.

Aster junciformis Rydb. Rush aster—Rare and local; sphagnum bogs; recently collected from a single station on Orcas Island. This constitutes one of the few records for Washington of this predominantly northern species. (FS)

Aster sibiricus L. var. *meritus* Arctic aster—Rare and local on the summit of Mt. Constitution. This northern species is only known from a few stations (e.g., Mt. Baker and the Olympic Mountains) in Washington. (FS)

Aster subspicatus Nees Douglas' aster—Uncommon and local, moist places near salt water.

**Bellis perennis* L. Lawn daisy—Common European lawn weed.

**Centaurea cyanus* L. Bachelor's button—Occasional weed; Friday Harbor.

**Centaurea diffusa* Lam. Tumble knapweed—Eurasian species, isolated at Lime Kiln Park, San Juan Island; rare. (SA)

**Centaurea maculosa* Lam. Spotted knapweed—Locally common weed in Friday Harbor; occasionally elsewhere.

**Centaurea pratensis* Thuill. Meadow knapweed—Towns and roadsides; locally common, especially on Lopez Island.

**Chondrilla juncea* L. Chondrilla—Eurasian adventive; reported as formerly present on Orcas Island by Baker.

**Chrysanthemum balsamita* L. Mint geranium—Occasional escape; Friday Harbor.

**Chrysanthemum leucanthemum* L. Oxeye-daisy—Abundant weed; roadsides and waste areas.

**Chrysanthemum maximum* Ramond. Shasta-daisy—Occasional escape to waste areas; frequently cultivated.

**Chrysanthemum parthenium* (L.) Bernh. Feverfew—Escaping and persisting; chiefly Friday Harbor; occasional.

**Cichorium intybus* L. Blue sailors—Common Eurasian adventive of roadsides and waste areas.

**Cirsium arvense* (L.) Scop. var. *horridum* Canada thistle—An aggressive, abundant weed; well-established in areas of prior disturbance; Eurasian.

Cirsium brevistylum Cronq. Short-styled thistle—Common; forest margins and roadsides.

**Cirsium vulgare* (Savi) Tenore Bull thistle—Disturbed sites; common throughout; Eurasian.

Conyza canadensis (L.) Cronq. var. *glabrata* Canadian fleabane—Occasional native weed of sandy waste places.

**Cotula coronopifolia* L. Brass buttons—Disturbed salt marshes; occasional; native of South Africa.

**Crepis capillaris* (L.) Wallr. Smooth hawksbeard—Abundant weed; roadsides and waste places throughout.

**Crepis nicaeensis* Balb. French hawksbeard—Occasional European weed, San Juan Island.

Crocidium multicaule Hook. Spring gold—Scattered and sporadic; rocky outcrops; mainly San Juan and Orcas islands.

Erigeron speciosus (Lindl.) DC. Showy fleabane—Rare; collected by Lyall on Lopez Island in 1858. More common east of the Cascades.

Erigeron trifidus Hook. Dwarf mountain daisy—Rare and local; summit of Mt. Constitution on steep rocky slopes. Probably a hybrid between *E. compositus* and *E. lanatus* (Packer, 1983). Listed in Hitchcock as *Erigeron compositus* var. *glabratus*.

Eriophyllum lanatum (Pursh) Forbes var. *lanatum* Common woolly-sunflower—Common, rocky outcrops to fairly disturbed meadows throughout.

**Filago germanica* L. German filago—Locally abundant throughout, from Cattle Point to Blakely and Stuart islands on gravelly, open disturbed sites; European.

Gnaphalium chilense Spreng. Cotton-batting plant—Occasional; dry forest margins to meadows; tolerant of disturbance. (WTU)

Gnaphalium microcephalum Nutt. var. *thermale* White cudweed—Dry, disturbed places; occasional. (WTU)

Gnaphalium palustre Nutt. Lowland cudweed—Dried mud puddles and open, moist disturbed places; common.

Gnaphalium purpureum L. var. *purpureum* Purple cudweed—Meadows and weedy sites; common.

**Gnaphalium uliginosum* L. Marsh cudweed—Disturbed sites; common European adventive.

Grindelia integrifolia DC. var. *macrophylla* Puget Sound gumweed—Abundant along rocky or sandy shorelines; occasionally extending into higher elevations on rocky substraits, as on Mt. Constitution.

Hieracium albiflorum Hook. White-flowered hawkweed—Common throughout; open woods to lichen balds.

**Hieracium pilosella* L. Mouse-ear hawkweed—Cultivated and marginally escaping/spreading in Friday Harbor.

**Hypochaeris glabra* L. Smooth cat's ear—Locally abundant; sandy meadows, San Juan Island; less common, but widespread, elsewhere; European.

**Hypochaeris radicata* L. Hairy cat's ear—European weed; abundant throughout.

Jaumea carnosa (Less.) Gray Jaumea—Fairly common; salt marshes; generally forms dense colonies.

**Lactuca ludoviciana* (Nutt.) Riddell Western lettuce—Rare; collected by M. Denton in 1974 near False Bay, San Juan Island. Not generally known from the Puget Sound area. (WTU)

**Lactuca muralis* (L.) Fresen. Wall lettuce—Abundant, woods; European.

**Lactuca serriola* L. Prickly lettuce—Common to abundant weed; waste sites throughout; European.

**Lapsana communis* L. Communist nipplewort—Aggressive Eurasian weed; common.

**Leontodon autumnalis* L. Fall hawkbit—Eurasian adventive; reported from San Juan Island.

**Leontodon nudicaulis* (L.) Merat Hairy hawkbit—Low, open disturbed sites, usually where moist; widespread but irregular.

Madia exigua (J.E. Smith) Gray Little tarweed—Fairly common; sloping grass prairies and rocky areas.

Madia glomerata Hook. Cluster tarweed—Occasional; disturbed to fairly pristine meadows. (FH)

Madia gracilis (J.E. Smith) Keck Slender tarweed—Fairly common and widespread; dry meadows.

Madia madioides (Nutt.) Greene Woodland tarweed—Common throughout; dry, open transitional forests.

Madia minima (Gray) Keck Small-head tarweed—Rare and local; on the summit of Mt. Constitution in open rocky prairies. Also reported from Waldron Island. Rather scarce in the Puget Sound area. (WTU)

Madia sativa Mol. Coast tarweed—Locally common on roadsides. Two forms occur with us; var. *congesta* and *sativa*; the latter more widespread, especially on San Juan Island.

**Matricaria maritima* L. Scentless May-weed—Occasional weed of disturbed, sandy places and strand; European.

Matricaria matricarioides (Less.) Porter Pineapple weed—Hard-packed waste soils, cracks in pavement, etc.; common native weed.

Microseris bigelovii (Gray) Schultz-Bip. Coast microseris—Rare and local; several widespread records, but now apparently restricted to San Juan Island. Prefers open, moist dunes or damp, shaded coastal bluffs; restricted in the Puget Sound area to the San Juan islands and Vancouver Island; more common in California. (WTU)

Petasites frigidus (L.) Fries var. *palmatus* Sweet coltsfoot—Rare; present at single station near Friday Harbor. (SA)

Psilocarphus tenellus Nutt. Slender woolly heads—Rare; vernal pools; collected by Dr. Adolf Ceska near Cattle Point. This is apparently the first Washington record for this California taxon. (V)

Senecio indecorus Greene Rayless mountain butterweed—Rare; Summit Lake, Orcas Island, and from a single station on San Juan Island. Prefers sphagnum bogs and lake edges; rare in Washington, common farther north. (WTU)

**Senecio jacobaea* L. Tansy ragwort—Fairly common Eurasian adventive along forest margins, usually where moist; becoming a nuisance in some areas.

Senecio macounii Greene Puget butterweed—Locally common; Mt. Constitution; open, rocky slopes and dry forest margins. Also reported from Blakely Island. A regional endemic. (WTU)

**Senecio sylvaticus* L. Wood groundsel—Occasional Eurasian adventive; dry, gravelly or rocky sites.

**Senecio vulgaris* L. Old-man-in-the-spring—Common introduced weed of roadsides and waste places.

Solidago canadensis L. var. *salebrosa* Canadian goldenrod—Fairly common species of roadsides, strand, and fields.

**Sonchus arvensis* L. Field milk-thistle—Locally common and widespread; landward edge of salt marshes and other low moist sites.

**Sonchus asper* (L.) Hill Prickly sow-thistle—Common European weed.

**Sonchus oleraceus* L. Common sow-thistle—Common European weed.

**Tanacetum vulgare* L. Common tansy—Occasional Eurasian weed; disturbed meadows and roadsides; most common on Lopez Island.

**Taraxacum laevigatum* (Willd.) DC. Red-seeded dandelion—Occasional; open rocky outcrops to waste places.

**Taraxacum officinale* Weber Common dandelion—Common weed; meadows to disturbed sites.

**Tragopogon dubius* Scop. Yellow salsify—Cultivated and occasionally escaping, especially on meadows; Orcas and San Juan islands.

**Tragopogon porrifolius* L. Salsify—Widespread adventive of roadsides and waste places; European.

Alismataceae

Alisma plantago-aquatica L. var. *americanum* American water-plantain—Localized; Trout Lake, San Juan Island; aquatic.

Hydrocharitaceae

Elodea canadensis Rich. in Michx. Canadian waterweed—Aquatic; occasional in lakes, often in deep water; most common on Orcas Island.

**Elodea densa* (Planch.) Casp. S. American waterweed—Aquatic; often used in aquariums and sometimes inadvertently introduced; well-established in a pond near Turn Point, San Juan Island.

Juncaginaceae

Triglochin concinnum Davy var. *concinnum* Graceful arrow-grass—Salt or brackish marshes; uncommon but widespread. (WTU)

Triglochin maritimum L. Seaside arrow-grass—Salt marshes; common.

Najadaceae

Najas flexilis (Willd.) Rost. & Schmidt Wavy water-nymph—Aquatic of shallow, muddy bottoms; rare; Spencer and Horseshoe lakes, Blakely Island.

Potamogetonaceae

Potamogeton amplifolius Tuckerman Large-leaved pondweed—Occasional; deep water of Cascade and Mountain lakes, Orcas Island, and Sportsman and Egg lakes, San Juan Island.

Potamogeton berchtoldii Fieb. Berchtold's pondweed—Shallow farm ponds to deep lakes; common throughout.

Potamogeton epihydrus Raf. Ribbon-leaved pondweed—Lakes on San Juan Island; rare.

Potamogeton foliosus Raf. var. *foliosus* Close-leaved pondweed—Shallow water of lakes and ponds; fairly common.

Potamogeton gramineus L. Grass-leaved pondweed—Shallow to medium-depth lakes; occasional; most common on San Juan Island. (WWB)

Potamogeton natans L. Broad-leaved pondweed—Farm ponds to larger lakes; common to abundant.

Potamogeton praelongus Wulf. White-stalked pondweed—Medium to deep water of lakes; widespread and common.

Potamogeton pusillus L. Small pondweed—Temporary ponds and shallow, muddy lakes; occasional; often growing with *P. berchtoldii* or *P. foliosus* var. *foliosus.* (WTU)
Potamogeton robbinsii Oakes Robbins' pondweed—Deep lakes, rather rare. (WWB)
Potamogeton zosteriformis Fern. Eel-grass pondweed—Lakes, often where fairly deep; common.

Ruppiaceae

Ruppia maritima L. Ditch-grass—Brackish ponds; occasional. (WTU)

Zosteraceae

Phyllospadix scouleri Hook. Scouler's surf-grass—Fairly common intertidal species; rocky bottoms. (FH)
Phyllospadix torreyi Wats. Torrey's surf-grass—Status unknown; collected near Eagle Cove, San Juan Island. A marine species. (FH)
**Zostera japonica* Japanese eel-grass—Rare; collected at Wescott Bay, San Juan Island. (FH)
Zostera marina L. Grass-wrack—Locally common intertidal species of protected, shallow bays.
**Zostera nana* Roth Dwarf eel-grass—Rare; collected at Cresent Bay, Orcas Island. (FH)

Juncaceae

Juncus acuminatus Michx. Tapered rush—Rather common, often occurring on the muddy fringe of grazed ponds and other wet places.
Juncus alpinus Vill. Northern rush—Rare; collected in 1964 by L.M. Sundquist on a muddy shoreline of Mountain Lake, Orcas Island. (WWB)
Juncus articulatus L. Jointed rush—Fairly common; ditches and other wet places.
Juncus balticus Willd. var. *balticus* Baltic rush—Widespread throughout the archipelago, coastal marshes to bogs near the summit of Mt. Constitution; very common.
Juncus bolanderi Engelm. Bolander's rush—Rare; collected by Peck three miles southeast of Roche Harbor. (WTU)
Juncus bufonius L. Toad rush—A common weed, from wet mucky areas to dirt roads.
Juncus effusus L. Common rush—Abundant throughout; farm fields, ditches, disturbed woodlands, etc. Three vars., *compactus, gracilis* and *pacificus* are represented in our material.
Juncus ensifolius Wikst. var. *ensifolius* Dagger-leaf rush—A common species of low moist ground and roadside ditches.
Juncus gerardii Loisel Mud rush—Uncommon; salt to brackish marshes, chiefly along the archipelago's southern edge, from Henry Island to Lopez Island. (WTU)
Juncus lesueurii Boland. Salt rush—Rare on strand; Cattle Point. Reported from Henry Island also. (WWB)
Juncus tenuis Willd. var. *tenuis* Slender rush—Rather common and somewhat weedy; pond margins, shaded dirt roads, etc.
Luzula campestris (L.) DC. Field woodrush—Abundant; outcrops and grassy meadows. A variable species but all our material appears to conform to var. *congesta.*

Cyperaceae

Carex aperta Boott Columbia sedge—Rare; moist areas. (WWB)
Carex aquatilis Wahl. Water sedge—Local on Mt. Constitution; sedge fens and lake margins.
Carex arcta Boott Northern clustered sedge—Swampy ground, mostly on Orcas Island, where sporadic.
Carex athrostachya Olney. Slender-beaked sedge—Wet meadows; occasional.
Carex aurea Nutt. Golden-fruited sedge—Quite common in flooded meadows, often where disturbed.
Carex brevicaulis Mack. Short-stemmed sedge—Rare; dunes; collected by Dr. A. Ceska on Cattle Point in May, 1983. (V)
Carex cusickii Mack. Cusick's sedge—Very common throughout; lake margins, bogs, sedge fens.
Carex deweyana Schw. Dewey's sedge—Moist coniferous and deciduous woodlands; exceedingly common.
Carex hendersonii Bailey Henderson's sedge—Uncommon in moist alder bottoms and mixed forests.

Carex hoodii Boott Hood's sedge—Locally common on Mt. Constitution, where it blankets open meadows in the vicinity of the Little Summit fire lookout tower. Rare elsewhere.
Carex interior Bailey Inland sedge—Reported by L. Kunze from Orcas Island; wet places.
Carex lasiocarpa Ehrh. Slender sedge—Occasional on Orcas Island in sphagnum bogs and fens.
Carex lenticularis Michx. var. *limnophila*—Rare in salt marshes and seeps along shorelines. (FS)
Carex leporina L. Hare sedge—Rare; collected by Henderson at East Sound in July, 1892. (WTU)
Carex leptalea Wahl. Bristle-stalked sedge—Fairly common on Orcas Island; generally on sphagnum.
Carex lyngbyei Hornem. Lyngby's sedge—Widespread but uncommon in its limited habitat of salt and brackish marshes.
Carex macrocephala Willd. Big-headed sedge—Locally common on strand, e.g., Spencer Spit, Lopez Island, Cattle Point, etc.
Carex muricata L. Muricate sedge—Occasional on low, wet ground.
Carex obnupta Bailey Slough sedge—Our most abundant sedge; occurring throughout in wet places. Often found in low, muddy sites in the forest where nothing else grows.
Carex pachystachya Cham. Thick-headed sedge—An occasional species of variable habitats, wet to dry areas, sometimes where disturbed.
Carex pansa Bailey Sand-dune sedge—Rare; collected by Dr. A. Ceska at Cattle Point in May, 1983. (V)
Carex pauciflora Lightf. Few-flowered sedge—Rare and local; bogs on Mt. Constitution. (WTU)
Carex pensylvanica L. var. *vespertina* Pennsylvania sedge—Abundant throughout on open prairies and rocky areas. Often a co-dominant with *Festuca rubra.*
Carex phyllomanica W. Boott Coastal stellate sedge—Rare; wet bogs on Orcas Island.
Carex praticola Rydb. Meadow sedge—Uncommon to rare; wet, shady situations.
Carex rossii Boott Ross sedge—Occasional; dry, open woods.
Carex rostrata Stokes Beaked sedge—Fairly common; sedge fens; Mt. Constitution and vicinity.
Carex sitchensis Prescott Sitka sedge—Lake margins and sedge fens; common throughout, particularly on Orcas Island.
Carex stipata Muhl. Sawbeak sedge—Scattered and uncommon; moist spots in alder woodlands, seeps, and pond borders.
Carex tumulicola Mack. Foothill sedge—Uncommon to rare; dry open meadows.
Carex vesicaria L. var. *major* Inflated sedge—Most common on Mt. Constitution; sedge fens and lake margins; occasional elsewhere.
Dulichium arundinaceum (L.) Britt. Dulichium—Occasional; sphagnum bogs; Orcas Island.
Eleocharis palustris (L.) R. & S. Creeping spike-rush—Very common; around margins of grazed ponds and lakes.
Eriophorum chamissonis C.A. Mey. Chamisso's cotton-grass—Local on sphagnum mats.
Eriophorum gracile Koch Slender cotton-grass—Rare and local in sphagnum bogs; summit area of Mt. Constitution. (WTU) (WWB)
Rhynchospora alba (L.) Vahl White beakrush—Rare; collected by Henderson in 1892 at Summit Lake. (WTU)
Scirpus acutus Muhl. Viscid bulrush—Widespread; on the periphery of protected salt marshes, lake margins, and sphagnum bogs; fairly common.
Scirpus americanus Pers. American bulrush—Rare; wet sandy beaches; collected by Lawrence on Stuart Island in July, 1904. (WTU)
Scirpus cyperinus (L.) Kunth var. *brachypodus* Wool-grass—Rare; restricted in Washington to the northern end of the Puget Trough. In the San Juans, it occurs sparingly in swampy areas on Orcas Island, particularly in the Mt. Woolard district.
Scirpus maritimus L. var. *paludosus* Seacoast bulrush—Rare with us, due to the lack of estuarine habitats that it prefers.
Scirpus microcarpus Presl. Small-fruited bulrush—Fairly common in low, wet areas, often occurring with *Phalaris.*

Gramineae

Agropyron caninum (L.) Beauv. Bearded wheatgrass—Open meadows; rare; ssp. *majus* var. *latiglume* and var. *unilaterale* both present. (WTU)

*_Agropyron repens_ (L.) Beauv. Quack grass—Common Eurasian adventive of pastures, meadows, and sandy beaches.

*_Agrostis alba_ L. Creeping bentgrass—Very common European escape; represented by var. _alba_ in farm fields and roadsides and var. _palustris_ in wet situations.

Agrostis diegoensis Vasey Leafy bentgrass—Dry prairies; rather uncommon.

Agrostis exarata Trin. Spike bentgrass—Widespread perennial, although uncommon; represented by the three variations: vars. _exarata, minor,_ and _monolepsis._ Meadows and roadsides.

Agrostis microphylla Steud. Small-leaved bentgrass—Rare; collected by Lawrence (1904). "Gravelly soil by salt water" on Stuart and Johns islands. (WTU)

Agrostis scabra Willd. Tickle-grass—Sporadic; generally occurring in moist areas. Collected most frequently on Orcas Island.

*_Agrostis tenuis_ Sibth. Colonial bentgrass—Native of Eurasia; very common in disturbed sites, roadsides, etc.

*_Aira caryophyllea_ L. Silver hairgrass—Abundant throughout on sandy or rocky soils.

*_Aira praecox_ L. Little hairgrass—Abundant; generally where rocky or gravelly.

Alopecurus aequalis Sobol. Little meadow foxtail—Locally common on the muddy fringe of farm ponds and small lakes; particularly numerous on San Juan Island.

*_Alopecurus geniculatus_ L. Water foxtail—Occasional in shallow water and wet fields.

*_Alopecurus pratensis_ L. Meadow foxtail—An introduced pasture species; occasional; moist meadows.

*_Anthoxanthum odoratum_ L. Sweet vernal grass—Roadsides and rocky meadows; common throughout.

*_Arrhenatherum elatius_ L. Presl. Tall oatgrass—Open meadows; generally where disturbed. Fairly common.

*_Avena fatua_ L. Wild oats—Occasional along roadsides and fields.

*_Avena sativa_ L. Cultivated oats—Escapee from cultivation and occasionally persisting for a few years in ditches and old fields.

Bromus carinatus H. and A. California brome—A fairly common perennial of dry, grassy meadows.

*_Bromus commutatus_ Schrad. Hairy chess—Occasional weed along roadsides; European.

*_Bromus inermis_ Leys. Hungarian brome—Sporadic weed of fields and roadsides.

*_Bromus mollis_ L. Soft brome—Undisturbed meadows to pastures and waste areas; abundant.

Bromus pacificus Shear Pacific brome—An uncommon native perennial of open, moist ground at lower elevations.

*_Bromus rigidus_ Roth Ripgut—Common weed; often grows on rocky sites.

Bromus sitchensis Trin. var. _sitchensis_ Alaska brome—Open rocky meadows and gravelly prairies; rather common.

*_Bromus sterilis_ L. Barren brome—European adventive; roadsides and waste places, sporadic.

*_Bromus tectorum_ L. Cheat grass—Abundant weed; south-facing meadows, rocky outcrops, and waste areas.

Bromus vulgaris (Hook.) Shear var. _vulgaris_ (Hook.) Columbia brome—Fairly common perennial of open woodlands and shady ravines.

Calamagrostis canadensis (Michx.) Beauv. var. _canadensis_ Bluejoint reedgrass—Occurring locally around lake margins; uncommon. Mostly on Orcas Island.

Cinna latifolia (Trevir.) Griseb. Woodreed—Occasional; riparian habitats.

*_Cynosurus cristatus_ L. Crested dogtail—Common weed of roadsides and meadows; European.

*_Cynosurus echinatus_ L. Hedgehog dogtail—Uncommon; roadsides and waste places; European.

*_Dactylis glomerata_ L. Orchard grass—Meadows, pasture land, and roadsides; very common.

*_Danthonia californica_ Boland. California oatgrass—Irregularly distributed, although occasionally becoming a dominant constituent of the native grass prairie.

Danthonia spicata (L.) Beauv. var. _pinetorum_ Common wild oatgrass—Dry transitional woodlands; uncommon.

Deschampsia cespitosa (L.) Beauv. Tufted hairgrass—Local in salt marshes and low, wet places; uncommon.

Deschampsia elongata (Hook.) Munro Slender hairgrass—Fairly common perennial of forest margins and disturbed areas.

Distichlis spicata (L.) Greene var. *borealis* Seashore saltgrass—Locally common along shorelines; more common in salt marshes.

**Echinochloa crus-galli* (L.) Beauv. Barnyard grass—Occasional weed of lots and roadsides.

Elymus glaucus Buckl. Blue wild rye—An abundant perennial of open grassy hillsides and transitional woodlands. Our material is represented by the vars. *glaucus* and *jepsonii.*

Elymus mollis Trin. Beach ryegrass—Common sand binder along beaches.

**Festuca arundinacea* Schreb. Reed fescue—A common, introduced pasture species escaping along roadsides, farm fields, and less disturbed meadows.

**Festuca bromoides* L. Six-weeks fescue—Widely distributed along rocky outcrops and roadsides; fairly common.

Festuca idahoensis Elmer var. *idahoensis* Idaho fescue—Uncommon and local; upland areas of Mt. Constitution, where it is collected with greater frequency than *F. rubra;* rare elsewhere. Open prairies.

Festuca megalura Nutt. Foxtail fescue—Common annual of dry, open areas.

Festuca microstachys Nutt. Nuttall's fescue—Weedy and common throughout.

**Festuca myuros* L. Rat-tail fescue—Yet another small, weedy fescue; common.

Festuca occidentalis Hook. Western fescue—Very common; typical of shady coniferous woodlands throughout.

Festuca ovina L. var. *rydbergii* Sheep fescue—Widespread but thinly distributed; rocky benchland and dry meadows.

Festuca rubra L. Trin. Red fescue—An abundant native perennial; often the primary constituent of the native grass prairie. Our material is represented by the vars. *littoralis* and *rubra.* Rocky benchland, meadows, and shorelines.

Festuca subulata Trin. Nodding fescue—Uncommon but widespread in dry to rather moist coniferous woodlands.

Festuca subuliflora Scribn. Coast range fescue—Coniferous woodlands throughout; fairly common.

Glyceria borealis (Nash) Batch. Northern mannagrass—Uncommon around the margins of lakes; Orcas and San Juan islands.

Glyceria occidentalis (Piper) Nels. Western mannagrass—Low wet situations and standing water; occasional.

**Holcus lanatus* L. Yorkshire fog—Abundant weed of open waste areas and fields to less disturbed sites.

Hordeum brachyantherum Nevski Meadow barley—Rather common in its limited habitat along salt marshes and sandy beaches.

**Hordeum geniculatum* All. Mediterranean barley—Occasional weed of fields and roadsides.

**Hordeum jubatum* L. Squirrel-tail—Uncommon; mostly in salt marshes but sometimes at waste places.

**Hordeum leporinum* Link. Charming barley—Disturbed sites; occasional weed from Europe.

**Hordeum murinum* L. Mouse barley—Fairly common weed of vacant lots and roadsides; European.

Koeleria cristata Pers. Prairie junegrass—Found growing in open conditions wherever it is dry and windy during the growing season; usually in undisturbed situations; common. Generally east of the Cascade Range.

**Lolium multiflorum* Lam. Italian ryegrass—Common weed; European.

**Lolium perenne* L. English ryegrass—Very common weed of fields and roadsides.

Melica subulata (Griseb.) Schribn. Alaska oniongrass—Typical of conifer woodlands; common.

**Phalaris arundinacea* L. Reed canarygrass—Often the dominant species of disturbed freshwater wetlands where it may cover extensive tracts; locally abundant. Also common in roadside ditches.

**Phleum pratense* L. Timothy—Fairly common roadside and pasture grass; European.

**Poa annua* L. Annual bluegrass—Lawns, roadsides, old rotting docks, etc.; common.

**Poa bulbosa* L. Bulbous bluegrass—Dry, gravelly disturbed sites; fairly common.

**Poa compressa* L. Canadian bluegrass—Fairly common and widespread; light woods and meadows, to the drier, upper parts of salt marshes.

Poa confinis Vasey Coastline bluegrass—Coastal meadows and dunes; locally common, from Cattle Point north to Stuart and Sucia islands.

Poa howellii Vasey & Scribn. Howell's bluegrass—Showing a propensity for steep banks in the shade of Douglas-fir, this native annual is seldom collected in the archipelago, but is probably more widespread than our records indicate. (WTU)

Poa pratensis L. Kentucky bluegrass—Abundant in several habitats; meadows, open forests, waste areas, etc. Often misidentified due to its variability.

Poa scabrella (Thurb.) Benth. Pine bluegrass—Widespread on rocky benchland throughout the archipelago; fairly common. Generally found east of the Cascade Range.

**Poa trivialis* L. Roughstock bluegrass—Fairly common; native of Europe; low, moist areas.

**Polypogon monspeliensis* (L.) Desf. Rabbitfoot—Moist, sandy areas adjacent to salt water; fairly common; European.

Puccinellia cusickii Weath. Cusick's alkaligrass—Reported by Kunze from three islands; salt marshes.

Puccinellia lucida Fern. & Weath. Shining alkaligrass—Reported by Kunze; salt marshes, Lopez Island.

Puccinellia maritima (Huds.) Parl. Coast alkaligrass—Occasional in salt marshes.

Puccinellia nutkaensis (Presl.) Fern. & Weath. Pacific alkaligrass—Rather rare; salt marshes. (WTU) (WWB)

Puccinellia nuttalliana (Schult.) Hitchc. Nuttall's alkaligrass—Rocky shores, salt marshes, and brackish areas; common.

Puccinellia pauciflora (Presl.) Munz var. *pauciflora* Marsh alkaligrass—Standing water around the periphery of lakes and ponds; fairly common.

**Setaria lutescens* (Weigel) Hubb. Yellow bristlegrass—Occasional weed, San Juan Island; collected by Baker in 1983.

Stipa lemmonii (Vasey) Scribn. var. *lemmonii* Lemmon's needlegrass—Gravelly meadows and rocky benchland; quite common.

Stipa occidentalis Thurb. var. *minor* Western needle-and-thread—Sporadically occurring on hilly prairies.

Trisetum canescens Buckl. Tall trisetum—Open transitional woodlands; fairly common.

Trisetum cernuum Trin. Nodding trisetum—Common in moist coniferous woodlands.

**Triticum aestivum* L. Wheat—Occasionally persisting in old fields and escaping along roadsides.

Sparganiaceae

Sparganium emersum Rehmann var. *emersum* Bur-reed—Common aquatic, San Juan Island, less common on other major islands.

Typhaceae

Typha latifolia L. Cat-tail—Abundant; fresh water.

Araceae

Lysichitum americanum Hulten & St. John Skunk cabbage—Damp valley bottoms and shaded fens, in Humic Gleyed soils; common.

Lemnaceae

Lemna minor L. Duckweed—Abundant floating aquatic.

Lemna trisulca L. Star duckweed—Small ponds and lake margins; uncommon.

Spirodela polyrhiza (L.) Schleid. Great duckweed—Shallow, boggy lakes; Blakely and Orcas islands.

Liliaceae

Allium acuminatum Hook. Hooker's onion—Dry gravelly meadows and rocky benchland; large and small islands; abundant.

Allium amplectens Torr. Slim-leaf onion—Rare, no recent sightings; known from a few old collections and reports. Argyle district on San Juan Island. Meadows by the sea. (WS)

Allium cernuum Roth Nodding onion—Widespread; rocky outcrops on the summit of Mt. Constitution to transitional forests, gravelly meadows and occasionally on strand.

**Asparagus officinalis* L. Asparagus—Occasional weed; often cultivated; strand and roadsides, mostly on Lopez Island.

Brodiaea coronaria (Salisb.) Engl. Harvest brodiaea—Locally abundant; dry meadows.

Brodiaea howellii Wats. Howell's brodiaea—Principally on San Juan Island in Mt. Dallas-Cady Mountain district; scattered on prairies to the south, and rare on Orcas Island. (FS)

Brodiaea hyacinthina (Lindl.) Baker Hyacinth brodiaea—Fairly common on dry, grassy prairies; San Juan Island, middle elevations on Mt. Constitution, etc.

Camassia leichtlinii (Baker) Wats. var. *suksdorfii* Great camas—Meadows throughout; abundant. A white form (var. *leichtlinii?)* with cleistogamous flowers is occasionally encountered in large populations of var. *suksdorfii.*

Camassia quamash (Pursh) Greene var. *azurea* Purple camas—Locally common on Mt. Dallas; scattered but widespread elsewhere.

Disporum hookeri (Torr.) Nicholson var. *oreganum* Hooker's fairybell—Rare; collected by F.H. Frye in 1908 near Mt. Dallas; moist woods. (FH)

Erythronium oreganum Applegate White fawn-lily—Dry coniferous forest margins to open meadows; fairly common, especially where undisturbed.

Fritillaria camschatcensis (L.) Ker-Gawl. Kamchatka fritillary—Rare; grassy bluffs by the sea; originally collected by Captain Archibald Menzies on Orcas Island in 1792; recently reported from Bald Hill, San Juan Island (Baker) and from Orcas Island; formerly present also near Cattle Point. Rare and near its southern maximum range in Washington.

Fritillaria lanceolata Pursh Chocolate lily—Meadows and open woods throughout; most common on coastal bluffs. Populations with pure yellow tepals occur at Iceberg Point, Lopez Island.

Lilium columbianum Hanson Columbia lily—Uncommon but widespread; dry open woodlands to gravelly, scrubby roadsides.

Maianthemum dilatatum (Wood) Nels. & Macbr. Wild lily-of-the-valley—Fairly common, although local; valleys, bluffs and woods; prefers moist, fertile soil.

**Scilla hispanica* L. Spanish bluebell—Introduced garden species; escaping and occasionally persisting in residential waste places.

Smilacina racemosa (L.) Desf. Giant Solomon's seal—Rare; Sucia and Lopez islands. Moist woods. (WWB)

Smilacina stellata (L.) Desf. Star-flowered Solomon's seal—Locally common in low, moist, partly shaded sites with fertile soil. Dense colonies are located at Davis Head, Lopez Island, the south end of Cascade Lake, Cattle Point, and elsewhere.

Zigadenus venenosus Wats. var. *venenosus* Poison-camas—Abundant; moist meadows throughout.

Iridaceae

**Iris germanica* L. Blue flag—Commonly cultivated, often spreading and long-persistent; also apparently escaped near Argyle Lagoon, San Juan Island.

**Iris pseudacorus* L. Yellow flag—Introduced and spreading along ponds and lakes; naturalized at a few places on San Juan and Orcas islands.

Sisyrinchium angustifolium Mill. Blue-eyed grass—Moist upland meadows; common to locally abundant.

Sisyrinchium californicum (Ker-Gawl.) Dryand. Golden-eyed grass—Rare; boggy lakeshores; collected by Peck at Cascade Lake in June, 1923. (WTU)

Sisyrinchium douglasii A. Dietr. Grass-widows—Moist meadows; occasional. (WTU)

Orchidaceae

Calypso bulbosa (L.) Oakes Calypso—Shady, dry to moist coniferous woods; fairly common to locally abundant, particularly in first- or second-growth forests.

Corallorhiza maculata Raf. ssp. *maculata* Spotted coralroot—Dry to moist shady coniferous forests; common throughout.

Corallorhiza mertensiana Bong. Merten's coralroot—Rare; dense, moist coniferous woods. Originally collected by Henderson on Mt. Constitution in 1892; since relocated there and reported from Mt. Dallas.

Corallorhiza striata Lindl. Striped coralroot—Dry Douglas-fir forests and margins; uncommon but widespread.

Eburophyton austiniae (Gray) Heller Phantom-orchid—Rare; dense Douglas-fir forests near

Entrance Mountain, Orcas Island. Formerly considered sensitive in Washington by the Washington Natural Heritage Program. Near its northern maximum range in the San Juans.

**Epipactis helleborine* (L.) Crantz European helleborine—European adventive, apparently spreading from Victoria; well-established at a number of places on San Juan Island, preferring gravelly roadsides to moist forest margins. A report of *E. gigantea*, a rare native in Washington, from Kanaka Bay is almost surely a misidentification of *E. helleborine*, found in that area.

Goodyera oblongifolia Raf. Rattlesnake-plantain—Moist, mossy coniferous forest floors; common.

Habenaria dilatata (Pursh) Hook. var. *dilatata* White bog-orchid—Rare; sphagnum bogs and lakeshores; Mt. Constitution and reported from San Juan Island. (WTU)

Habenaria elegans (Lindl.) Boland. Elegant rein-orchid—Dry, open coniferous forests; common throughout.

Habenaria greenei Jeps. Greene's rein-orchid—Dry coniferous forest margins, dry meadows, and even strand; widespread and fairly common on all larger and many smaller islands. Mostly confined to the San Juans in western Washington; more common in California.

Habenaria orbiculata (Pursh) Torr. Round-leaved rein-orchid—Rare; dry, open coniferous woods; collected by Foster near Friday Harbor in 1904. (WTU)

Habenaria unalascensis (Spreng.) Wats. Unalaska rein-orchid—Gravelly, dry forests to fairly open places; sporadic but fairly common throughout.

Liparis loeselii (L.) L.C. Rich. Twayblade—Rare; sphagnum bogs on the west lobe of Orcas Island, where it grows with associates such as *Ophioglossum vulgatum, Lycopus uniflorus, Menyanthes trifoliata,* and *Agrostis scabra.* Formerly only known in Washington (and in the entire Pacific Northwest) from Skamania County, where recently rediscovered. Described as endangered for our state by the Washington Natural Heritage Program; more common east of the Rocky Mountains. (FS)

Listera caurina Piper Western twayblade—Deep, moist coniferous woods; rare; San Juan and Orcas islands. (WTU)

Listera cordata (L.) R. Br. Heart-leaved twayblade—Deep, mossy coniferous forest floors; uncommon and irregular. (WTU)

Spiranthes romanzoffiana Cham. var. *romanzoffiana* Hooded lady's tresses—Vernal pools and moist meadows; fairly common throughout.

BIBLIOGRAPHY

Adams, R.P. Intraspecific Terpenoid Variation in *Juniperus scopulorum*: Evidence for Pleistocene Refugia and Recolonization in Western North America. *Taxon* 32:31-46. 1983.

Armstrong, J.E., Crandell, I.R., Easterbrook, D.R., and Nobel, J.B. Last Pleistocene Stratigraphy and Chronology in Southwestern British Columbia and Northwestern Washington. *Geological Society of America Bulletin* 76:321-30. 1965.

Atkinson, S.R., and Sharpe, F.A. Ferns of San Juan County. *Douglasia* 7:2-4. 1983.

Brandon, M.T., Cowan, D.S., Muller, J.E., and Vance, J.A. Pre-Tertiary Geology of the San Juan Islands, Washington and Southeast Vancouver Island, British Columbia. *Geological Association of Canada.* May, 1983.

Buckingham, N.M., and Tisch, E.L. *Vascular Plants of the Olympic Peninsula, Washington.* Cooperative Park Studies Unit, National Park Service CPSU/UW B-79-2. College of Forest Resources, University of Washington, Seattle. 1979.

____________________. Additions to the Native Vascular Flora of the Olympic Peninsula, Washington. *Madrono* 30:67-78. 1983.

Ceska, A. and O. Additions to the Flora of British Columbia. *Canadian Field-Naturalist* 94:69-74. 1980.

Denton, M.F. *Ranunculus californicus*, A New Record for the State of Washington. *Madrono* 25:132. 1977.

Detling, L.E. Historical Background of the Flora of the Pacific Northwest. *Museum of Natural History Bulletin* No. 13. University of Oregon, Eugene. 1968.

Dodd, V.E., and Tusler, H.G. *Wild Flowering Plants on Coon, McConnell, Reef, and Yellow Islands.* Unpublished report. 1958.

Dunwiddie, P.W. *Yellow Island Vegetation Plot Study.* Unpublished report. 1982.

Easterbrook, D.J. Late Pleistocene Glacial Events and Relative Sea Level Changes in the Northern Puget Lowland, Washington. *Geological Society of America Bulletin* 74:1465-84. 1963.

Eaton, C.M. *Management Plan and Baseline Study for Goose and Deadman Island Preserves, San Juan County, Washington.* University of Washington, Seattle. 1978.

Fonda, R.W., and Bernardi, J.A. Vegetation of Sucia Island in Puget Sound, Washington. *Bulletin of the Torrey Botanical Club* 103:3:99-109. 1976.

Franklin, J.F., and Dyrness, C.T. *Natural Vegetation of Oregon and Washington.* U.S. Dept. of Agric. Forest Ser. Gen. Tech. Report PNLU-8. 1973.

Frye, T.C. *Ferns of the Northwest.* Metropolitan Press, Portland, Oregon. 1934.

Hansen, H.P. A Pollen Study of Two Bogs on Orcas Island of the San Juan Islands, Washington. *Bulletin of the Torrey Botanical Club* 70:3:236-43. 1943.

Henry, J.K. *Flora of Southern British Columbia and Vancouver Island.* W.J. Gage and Co., Ltd., Toronto. 1915.

Hitchcock, C.L., and Cronquist, A. *Flora of the Pacific Northwest.* University of Washington, Seattle. 1974.

Hitchcock, C.L., Cronquist, A., Ownby, M., and Thompson, J.W. *Vascular Plants of the Pacific Northwest.* University of Washington, Seattle. 1955-69.

Huesser, C.J. Environmental Sequence following the Fraser Advance of the Juan de Fuca Lobe. *Quaternary Research* 3:284-306. 1973a.

Janzsen, H.V. Vascular Plants of Mayne Island, British Columbia. *Syesis.* 14:81-92. 1977.

______________. Vascular Plants of Saturna Island, British Columbia. *Syesis.* 10:85-96. 1981.

Jones, G.N. *A Botanical Survey of the Olympic Peninsula, Washington.* University of Washington Publications in Biology 5, Seattle. 1936.

Kozloff, E.N. *Plants and Animals of the Pacific Northwest.* University of Washington, Seattle. 1976.

Krajina, V.J. Ecology of Forest Trees in British Columbia. *Ecology of Western North America* 2:1-147. 1951.

Krajina, V.J., and Spilsbury, R.H. Forest Associations on the East Coast of Vancouver Island. *Forestry Club Handbook.* 1953.

Kruckeberg, A.R. Plant Life on Serpentinite and Other Ferromagnesian Rocks in Western North America. *Syesis.* 2:15-144. 1969.

______________. *Checklist of Seed Plants and Ferns Seen or Collected on Cypress Island.* University of Washington, Seattle. 1967.

McLelland, R.D. *Geology of the San Juan Islands.* University of Washington Publications in Geology 2. 1927.

Miller, J.P. *Washington Climate.* Washington State University, Pullman. 1966.

Mottola, M. and K. *Marsh-hill Wildlife Haven Plant List.* Unpublished list. 1974.

Newcomb, C.F. *Menzies' Journal of Vancouver's Voyage, April to October.* Archives of British Columbia. *Memoir* No. 5. Victoria, B.C. 1923.

Patterson, C. *Vegetation Study of the Waldron Reserve.* Report for The Nature Conservancy. Unpublished report. 1976.

Peterson, L.K., Mehringer, C.E., and Gustafson, C.E. Late-glacial Vegetation and Climate at the Manis Mastodon Site, Olympic Peninsula, Washington. *Quaternary Research* 20:215-231. 1983.

Phillips, E.L. *Washington Climate.* Coop. Ext. Ser. Wash. State Univ. Pullman. 1966.

Piper, C.V. *Flora of the State of Washington.* U.S. Government Printing Office. 1906.

Richardson, F. *Ecological Inventory of the Crowley-Syre Property.* Blakely Island, Washington. Unpublished report. 1974.

Rigg, G.B. "Forest Distribution in the San Juan Islands: A Preliminary Note." *Plant World* 16:177-182. 1913.

Russell, R.H. (Ed). "Geology and Water Resources of the San Juan Islands." Washington Department of Ecology *Water Supply Bull.* 46. Olympia. 1975.

Slater, J.R. *Distribution of Selected Taxa in Western Washington.* Occasional Papers in Biology, University of Puget Sound. 1974.

Soil Conservation Service. *Soil Survey. San Juan County, Washington.* U.S.D.A. Washington, D.C. 1962.

Szczawinski, A.F., and Harrison, A.S. *Flora of the Saanich Peninsula.* Occasional Papers, British Columbia Provincial Museum No. 20 Queens Printer, Victoria. 1973.

The Nature Conservancy. *San Juan County, Washington: Inventory of Natural Areas on Private Lands.* Olympia. 1975.

The Nature Conservancy. *Sentinel Island Survey.* Olympia. 1982.

Washington Natural Heritage Program. *An Illustrated Guide to the Endangered, Threatened and Sensitive Vascular Plants in Washington.* Olympia. 1981.

Washington Natural Heritage Program. *Endangered, Threatened and Sensitive Plants of Washington.* Olympia. 1984.

INDEX

This index contains common and Latin names found on pages 12 through 131. Bold-faced page numbers refer to illustrations.

About the authors:

FRED SHARPE, who received a B.S. in Botany from the University of Washington, has been engaged in field research on San Juans flora for more than four years. Sharpe produced the illustrations for this book and did most of the field work. A resident of Fidalgo Island, Washington, he is continuing his research on birds and plants of the Islands. He is a member of the Washington Native Plant Society and the Seattle Audubon Society.

SCOTT R. ATKINSON, who wrote the text of this book, holds a B.A. in Russian regional studies with a major in political science from the University of Washington, where he is now working for a master's degree. Atkinson was responsible for most of the herbarium research involved in the preparation of this field guide, and has been compiling data on San Juans flora for more than seven years. A Seattle resident, he is a member of both the Seattle and San Juans Audubon societies.